Applications of Cloud Computing

Chapman & Hall/CRC Distributed Sensing and Intelligent Systems Series
Series Editor: Mohamed Elhoseny and Xiaohui Yuan

Applications of Cloud Computing
Approaches and Practices
Prerna Sharma, Moolchand Sharma, Mohamed Elhoseny

Applications of Cloud Computing

Approaches and Practices

Edited By

Prerna Sharma
Department of Computer Science and Engineering,
Maharaja Agrasen Institute of Technology, GGSIPU, Delhi, India

Moolchand Sharma
Department of Computer Science and Engineering,
Maharaja Agrasen Institute of Technology, GGSIPU, Delhi, India

Mohamed Elhoseny
College of Computer Information Technology,
American University in the Emirates,
United Arabic Emirates

CRC Press
Taylor & Francis Group
Boca Raton London New York

CRC Press is an imprint of the
Taylor & Francis Group, an **informa** business
A CHAPMAN & HALL BOOK

First edition published 2021
by CRC Press
6000 Broken Sound Parkway NW, Suite 300, Boca Raton, FL 33487-2742

and by CRC Press
2 Park Square, Milton Park, Abingdon, Oxon, OX14 4RN

Library of Congress Cataloging-in-Publication Data

Names: Sharma, Prerna, editor. | Sharma, Moolchand, editor. | Elhoseny,
Mohamed, editor.
Title: Applications of cloud computing : approaches and practices / edited
by Prerna Sharma, Department of Computer Science & Engineering, Maharaja
Agrasen Institute of Technology, GGSIPU, Delhi, India, Moolchand Sharma,
Department of Computer Science & Engineering, Maharaja Agrasen Institute
of Technology, GGSIPU, Delhi, India, Mohamed Elhoseny, Faculty of
Computers and Information Mansoura University Dakahlia, Egypt.
Description: First edition. | Boca Raton : CRC Press, 2021. | Series:
Chapman & Hall/CRC distributed sensing and intelligent systems series |
Includes bibliographical references and index.
Identifiers: LCCN 2020030143 | ISBN 9780367904128 (hardback) | ISBN
9781003025696 (ebook)
Subjects: LCSH: Cloud computing.
Classification: LCC QA76.585 .A67 2021 | DDC 004.67/82--dc23
LC record available at https://lccn.loc.gov/2020030143

ISBN: 978-0-367-90412-8 (hbk)
ISBN: 978-1-003-02569-6 (ebk)

Typeset in Minion
by SPi Global, India

Prerna Sharma dedicates this book to her parents, Vipin and Suman Sharma, and her husband, Parminder Mann, for their constant support and motivation, and her mentor for his valuable guidance. Specially dedicated to her beloved son, Pratyaksh Mann.

Moolchand Sharma dedicates this book to his parents, Sh. Naresh Kumar Sharma and Smt. Rambati Sharma, for their constant support and motivation, and his mentor, Dr. Suman Deswal, for her valuable guidance, and also to his family, including his wife, Pratibha Sharma, and his son, Dhairya Sharma. Above all, thanks to Almighty God.

Dr. Mohamed Elhoseny dedicates this book to his parents, his wife, and his family for their valuable guidance and constant motivation.

Contents

Tables and Figures

Preface

WE ARE DELIGHTED TO LAUNCH *APPLICATIONS OF CLOUD COMPUTING: Approaches and Practices*. Cloud computing provides a cheap computing framework for a large volume of data for real-time applications. It is, therefore, not surprising that cloud computing has become a buzzword in the computing fraternity over the last decade. This book presents the various approaches, techniques, and applications that are available for cloud computing. It is a valuable source of knowledge for researchers, engineers, practitioners, and graduate and doctoral students working in the field of cloud computing. It will also be useful for faculty members of graduate schools and universities. Around 25 full-length chapters addressing a wide range of research areas were originally received, ten of which are included in this volume. All the chapters submitted were peer-reviewed by at least two independent reviewers, whose comments were communicated to the authors for incorporation in their revised manuscripts. The reviewers' recommendations were taken into consideration in the selection of chapters for inclusion in the volume. We believe that this exhaustive review process ensured that each published chapter met rigorous academic and scientific standards.

We would like to thank the authors of the published chapters for adhering to the schedule and for incorporating the review comments. We wish to extend our heartfelt appreciation to the authors, peer reviewers, committee members, and production staff whose diligent work enabled this volume to take shape. We especially want to thank the dedicated team of peer reviewers who volunteered for the arduous and tedious task of quality checking and critiquing the submitted chapters.

Prerna Sharma
Moolchand Sharma
Mohamed Elhoseny

Editors

Prerna Sharma is currently an assistant professor in the Department of Computer Science and Engineering at Maharaja Agrasen Institute of Technology, GGSIPU Delhi. She has been teaching for eight years and is a doctoral researcher at Delhi Technological University (DTU), Delhi. She graduated from GPMCE, GGSIPU in 2009 and completed her postgraduate studies at USIT, GGSIPU in 2011. She has authored/co-authored SCI-indexed journal and Scopus-indexed journal articles in prestigious journals such as *Journal of Supercomputing* (Springer), *Cognitive Systems Research* (Elsevier), *Expert Systems* (Wiley), and *International Journal of Innovative Computing and Applications* (Inderscience). She has also authored book chapters published by Wiley and Elsevier. She has worked extensively in the field of computational intelligence. Her areas of interest include artificial intelligence, machine learning, nature-inspired computing, soft computing, and cloud computing. She is associated with professional bodies including IAENG, ICSES, UACEE, and the Internet Society.

Moolchand Sharma is currently an assistant professor in the Department of Computer Science and Engineering at Maharaja Agrasen Institute of Technology, GGSIPU Delhi. He has been teaching for more than seven years and is a doctoral researcher at DCR University of Science and Technology, Haryana. He graduated in 2010 from KNGD MODI Engineering College, GBTU, and completed his postgraduate studies at SRM University, NCR Campus, Ghaziabad, in 2012. He has published scientific research papers in international journals and conference proceedings including SCI-indexed and Scopus-indexed journals such as *Cognitive Systems Research* (Elsevier), *International Journal of Image & Graphics* (World Scientific), *International Journal of Innovative Computing and Applications* (Inderscience) and *Innovative Computing and Communication Journal.* He is also the co-convener of the ICICC Springer conference series. He has authored/co-authored chapters published by Elsevier, Wiley, and De Gruyter. His research areas include artificial intelligence, nature-inspired computing, security in cloud computing, machine learning, and search engine optimization. He is associated with various professional bodies including IAENG, ICSES, UACEE, and the Internet Society.

Dr. Mohamed Elhoseny is an assistant professor at the Department of Computer Science, College of Computer & Information Technology, American University in the Emirates (AUE). Dr. Elhoseny is an ACM Distinguished Speaker and IEEE Senior Member. He received his PhD in Computers and Information from Mansoura University/University of North Texas through a joint scientific program. Dr. Elhoseny is the founder and the Editor-in-Chief of IJSSTA journal published by IGI Global. Also, he is an associate editor at *IEEE Journal of Biomedical and Health Informatics, IEEE Access, Scientific Reports, IEEE Future Directions, Remote Sensing, and International Journal of E-services and Mobile Applications*. He has served as co-chair, publication chair, program chair, and track chair for several international conferences published by recognized publishers such as IEEE and Springer. Dr. Elhoseny is the Editor-in-Chief of the Springer *Studies in Distributed Intelligence* book series, Editor-in-Chief of the *Sensors Communication for Urban Intelligence* and *Distributed Sensing and Intelligent Systems* book series from CRC Press/Taylor & Francis Group.

Contributors

Satwik Bhardwaj
Department of Computer Engineering
Netaji Subhas University of Technology
(formerly Netaji Subhas Institute of
Technology)
Delhi, India

Anjali Chaudhary
Department of Computer Science and
Engineering
Maharaja Agrasen Institute of Technology
Delhi, India

Rahul Chawla
Maharaja Agrasen Institute of Technology
Delhi, India

Krishna Choudhary
Maharaja Agrasen Institute of Technology
Delhi, India

Dipta Datta
Lovely Professional University
Phagwara, India

Namrata Dhanda
Amity University
Uttar Pradesh, India

Kavita Dhull
Lovely Professional University
Phagwara, India

Deepak Gupta
Maharaja Agrasen Institute of Technology
Delhi, India

Kalpana Gupta
Center for Development of Advanced
Computing
Noida, India

Kshitij Gupta
Maharaja Agrasen Institute of Technology
Delhi, India

Bramah Hazela
Amity University
Uttar Pradesh, India

Mansi Jain
Maharaja Agrasen Institute of Technology
Delhi, India

Rahul Johari
SWINGER: Security, Wireless, IoT Network
Group of Engineering and Research
University School of Information,
Communication and Technology (USICT)
and
Guru Gobind Singh Indraprastha University
Delhi, India

Parmita Kain
Maharaja Agrasen Institute of Technology
Delhi, India

Anubha Khanna
Department of Computer Science and
    Engineering
Maharaja Agrasen Institute of Technology
Delhi, India

Ashish Khanna
Maharaja Agrasen Institute of Technology
Delhi, India

Akanksha Kochhar
Department of Computer Science and
    Engineering
Maharaja Agrasen Institute of Technology
Delhi, India

Karuna Middha
Department of Computer Science and
    Engineering, MAIT
Maharaja Agrasen Institute of Technology
Delhi, India

Rani
Department of Computer Science &
    Engineering, DCRUST
Murthal, Haryana
Delhi, India

Joel J. P. C. Rodrigues
National Institute of Telecommunications
    (Inatel)
Santa Rita do Sapucaí, MG, Brazil

Srishti Sahni
Maharaja Agrasen Institute of Technology
Delhi, India

Samridhi Seth
SWINGER: Security, Wireless, IoT Network
    Group of Engineering and Research
University School of Information,
    Communication and Technology (USICT)
and
Guru Gobind Singh Indraprastha University
Delhi, India

Deepak Kumar Sharma
Department of Information Technology
Netaji Subhas University of Technology
    (formerly Netaji Subhas Institute of
    Technology)
Delhi, India

Nitigya Sharma
Department of Computer Science and
    Engineering
Maharaja Agrasen Institute of Technology
Delhi, India

Prayag Tiwari
University of Padua
Italy

Prastuti Upadhaya
Department of Computer Engineering
Netaji Subhas University of Technology
    (Formerly Netaji Subhas Institute of
    Technology)
Delhi, India

Sahil Verma
Lovely Professional University
Phagwara, India

About This Book

CLOUD COMPUTING HAS CREATED A SHIFT FROM PHYSICAL HARDWARE and locally managed software-enabled platforms to virtualized cloud-hosted services. Cloud assembles large networks of virtual services, including hardware (CPU, storage, and network) and software resources (databases, message queuing systems, monitoring systems, and load-balancers). As cloud continues to revolutionize applications in academia, industry, government, and many other fields, the transition to this efficient and flexible platform presents serious challenges at both theoretical and practical levels—ones that will often require new approaches and practices in all areas. The future of cloud computing is assured by its superior power, extreme agility, accessibility, reliability, security, and high performance, which enable organizations to conduct their business more affordably. While still in its initial development and growth phase, the cloud has the potential to meet and resolve its upcoming challenges.

This book lays the foundations of the core concepts and principles of cloud computing applications, taking the reader through the fundamental ideas with expert ease. It reinforces theory with a full-fledged pedagogy designed to enhance students' understanding and offer them practical insights into cloud computing applications. Some novel ideas in cloud computing are introduced, including edge computing, fog computing, security and privacy issues in cloud computing, edge-fog computing and cloud cryptography.

CHAPTER 1

Analysis of Biological Information Using Statistical Techniques in Cloud Computing

Srishti Sahni, Rani, Ashish Khanna,
and Joel J. P. C. Rodrigues

CONTENTS

1.1 WHAT IS BIOINFORMATICS?

To understand bioinformatics, one must first understand data and the form they take when dealing with biology. Data are nothing but a collection of raw facts and figures that are processed and analyzed to make sense of observed variations, patterns, and values. Biology, as a whole, deals with a variety of large chunk databases such as experimental observations, genomics, proteomics, metabolomics, microarray gene expression, and many more [1]. These data are of no use in medicine until processed and analyzed [1]. The genome analysis of a person can take days and still be incomplete if processed manually; even if the processing is completed it may not be 100% reliable as the possibility of human error will always remain. This is where bioinformatics comes in (Figure 1.1).

Bioinformatics is an interdisciplinary field of study that combines biology, computer science, information engineering, mathematics, and statistics to analyze and interpret biological data. It is often regarded as a part of computational biology, which is a quantitative analytical technique that is used in biological systems for problem solving and data modeling [3]. Bioinformatics is focused on the development of software tools and techniques that help analysts and medical personnel to manage, analyze, and manipulate extremely large volumes of biological data.

> Bioinformatics is an interdisciplinary field of study that enables the user to manage, analyze, manipulate, and make sense of biological data.

1.1.1 A Brief History

The term "bioinformatics" was first coined in the 1970s by Paulien Hogeweg and Ben Hesper. They described bioinformatics as the study of information processes in biotic systems, which is not what it means today as this definition placed it closer to biochemistry.

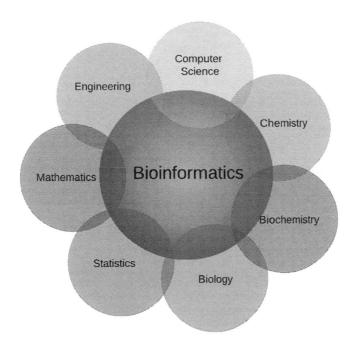

FIGURE 1.1   Bioinformatics: An interdisciplinary field.

The present definition and the field of bioinformatics itself did not gain momentum until the early 1990s. The use of computers in the field of molecular biology was introduced in the 1950s after Frederick Sanger determined the first sequential structure of a protein (insulin). It was next to impossible to compare multiple protein sequences manually, hence computers were introduced for sequence detection and analysis in the fields of microbiology for the very first time [2]. Margaret Oakley Dayhoff compiled the first protein sequence database and became a pioneer in this field. Her efforts, along with those of Elvin A. Kabat and Tai Tew, laid the groundwork for the field we today know as bioinformatics. This was followed by the formulation of The Protein Data Bank and the SWISSPROT protein sequence database in 1972 and 1987, respectively. The original databases were stored in flat files, either as a single large text file, or one per entry [6]. A huge variety of biological databases are now easily available in either public or commercialized third-party domains for use in the field of bioinformatics. After the formulation of biological databases, a wide variety of tools were made available in an attempt to search patterns and sequences from these databases. This began with simple keyword-matching algorithms, later progressing to sophisticated alignment-based pattern-matching methods [4]. The BLAST algorithm had been the mainstay of sequence database searching since its introduction a decade ago; this success and widespread acceptability is often attributed to its time efficiency [4]. BLAST is more rapid but less rigorous than its complementary FASTA and Smith Waterman algorithms. The following section describes the development and growth of bioinformatics since 2000.

Bioinformatics has come a long way since its introduction in the early 1990s and still has a long way to go (Figure 1.2.). Today it is at the heart of many types of research and ongoing projects. A lot of money is being invested in the collection, transfer, and exploitation of biological data, as it allows one to gain a better understanding of the living world [5].

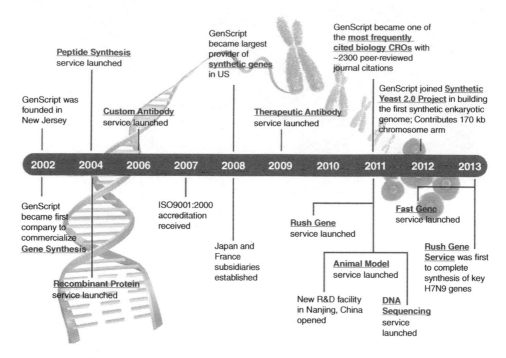

FIGURE 1.2 Developments in bioinformatics.

1.1.2 Tools and Techniques

The market is flooded with a large number of standard and customized software products that are used for the analysis of biological data. A large number of data-mining software and visualizing tools are available to retrieve data from genomic sequence databases, and to analyze and retrieve information from proteomic databases [7]. The following are among the tools used in bioinformatics.

- *Homology and similarity tools*: These tools are used to identify similarities between newly determined structures and per-examined existing structures, as they measure the divergence of the sequence from a common ancestor. The homology of two structures is determined as true or false based on the degree of similarities between them [8].

- *Protein function analysis*: These tools allow the user to estimate the biochemical functions of the protein in question. This is achieved by analyzing the motifs, signatures, and protein domains of the protein sequence in comparison with the secondary structures stored in the pre-formulated database [8].

- *Structural analysis:* These tools allow the user to compare the 2D/3D structures of the query with known biological structures. This is a widely used technique as the functions of a protein are based on their structures rather than sequences of their structural homologs [8].

- *Sequence analysis:* These sets of tools carry out a more detailed analysis of the query resulting in the identification of mutations, hydropathy regions, CpG islands, and compositional biases as well as evolutionary analysis [8].

The following are some examples of the tools listed above.

- *BLAST (Basic Local Alignment Search Tool)*: BLAST is an example of homology tools. These sets of tools are designed specifically for a Windows system and perform fast query searches regardless of the type of structure [9]. They are equipped for the comparison of the nucleotide sequences in a database. NCBI has recently introduced a new queuing system to BLAST that allows the users to format the results multiple times and retrieve them at their convenience.

There are a variety of programs based on the type of structures

1. blastp compares an amino acid query sequence against a protein sequence database.

2. blastn compares a nucleotide query sequence against a nucleotide sequence database.

3. blast compares a nucleotide query sequence translated in all reading frames against a protein sequence database.

4. tblastn compares a protein query sequence against a nucleotide sequence database dynamically translated in all reading frames.

5. tblastx compares the six-frame translations of a nucleotide query sequence against the six-frame translations of a nucleotide sequence database.

- *FASTA*: This is a homology program for protein sequences introduced in 1988 by Pearson and Lipman. A prescreen step is added to the algorithm that locates the highly matching sub-sequences/segments of the two sequences, which are further extended to their localities using algorithms like Smith–Waterman [10].

- *EMBOSS (European Molecular Biology Open Software Suite)*: It is a package that comes with a set of sequence-analysis programs, extensive libraries, and transparent web support, and is fully compatible with all UNIX/LINUX-based platforms [10].

- *Cluster*: It is a fully automated tool for DNA and protein sequencing that looks for the best match among the entire set of input sequences irrespective of its nature, i.e., whether it is a protein or amino acid [10].

1.1.3 Future Scope

Large biological databases are being formulated with every passing day, and active work is being done to improve and develop various techniques for analysis of these biological data-sets, making bioinformatics the center of attention for many research projects and ongoing developments in the USA, many European countries, and now also India [11]. It is continuously growing and, with the number of jobs constantly growing, it is in no danger of becoming overpopulated or crowded either in the near or distant future. Some of the biggest drug firms, such as Merck, Johnson & Johnson, Smith Kline Beecham, and

Glaxo Welcome, are constantly seeking bioinformatics experts, and the job market is expanding, in contrast to the opposite trend that can be observed in other sectors. This makes bioinformatics a viable and attractive choice for many young scientists and researchers.

1.2 CLOUD ANALYTICS

Cloud-based services have gained a good deal of momentum during the past decade. A number of cloud service models exist, each designed to serve different user computational needs; Cloud Analytics is one such service model [4]. A cloud-based analysis system delivers the elements of data analysis via a public or private cloud platform. Cloud analytics applications and services are often available under a utility (pay-per-use) or subscription-based pricing model.

The six critical elements of analytics are

- data sources

- data models

- processing applications

- computing power

- analytical models, and

- sharing and storage of results.

Any model which offers one or more of the above-listed services via a cloud platform is termed a cloud analytics model [4]. Various companies that offer only one such service often refer to themselves as cloud analytics-based companies—which is confusing for the users. There are various cloud analytic services offered by various tech giants in the business. Some of them are discussed in the following section.

1.2.1 Analytic Services

The cloud market is flooded with service providers offering various tools and services to the public according to their needs and demands. Microsoft's Azure, Amazon's AWS, and Google's cloud services are the major service providers in the cloud sector, and naturally they provide a variety of analytical services as well [12]. These services have impacted analytics as a whole, revolutionizing it to a great extent. Some of the available cloud-based analytic services are:

- *Hosted data warehouses*: A centralized repository that hosts the data for the users of the company. All the information is stored in a remote location and is accessed by users everywhere rather than being stored at the company's own systems [4].

- *Software as a Service business intelligence*: Offers support to various start-up ventures and delivers business intelligence applications from a hosted location. The services may not be as extensive as in-house applications, but they are easier to set up and are cost-effective [4].

- *Social media analytics*: These tools help the users choose from a variety of social media platforms that best serve the interests of their organization, allowing them to harvest and collect data from these platforms [1]. Various other data analytics tools are then used to make sense of the gathered information and draw conclusions.

- *Hosted data analytics*: These sets of software allow the user to analyze and process their data in a cost-effective manner [4].

A variety of data analytics software, along with hosted data warehouses, can be used to set up a cloud-based bioinformatics analysis system. These models will be discussed in further detail in the remainder of this chapter.

1.2.2 Bioinformatics Cloud

A bioinformatics cloud is a cloud-based delivery system that offers various web-based services to the user that facilitate the analysis and storage of bioinformatics data-sets.

A bioinformatics cloud offers a variety of analytical tools that can be put to use in bioinformatics data analysis. A large number of Data as a Service (DaaS), Software as a Service (SaaS), Platform as a Service (PaaS), and Infrastructure as a Service (IaaS) tools are available that fit the needs of the users in an error-free and cost-effective manner [9] (Figure 1.3.).

- *DaaS*: offers a dynamic virtual space hosted via web servers for the storage of the vast amounts of data used in bioinformatics data analysis. It also enables connected users to access the updated data in real time using a network of systems across the globe, and can prove fruitful during analysis [1].

- *SaaS*: offers multiple pre-existing cloud-based tools and bioinformatics applications that target mapping applications, sequences alignment, and gene expression analysis [1].

- *PaaS*: these solutions allow users to customize their own bioinformatics applications, unlike SaaS, enabling them to control the instances and data involved with each application [1].

- *IaaS*: a virtual system offering computational benefits. The user has full control over the deployed storage resources, operating systems, and bioinformatics applications [1].

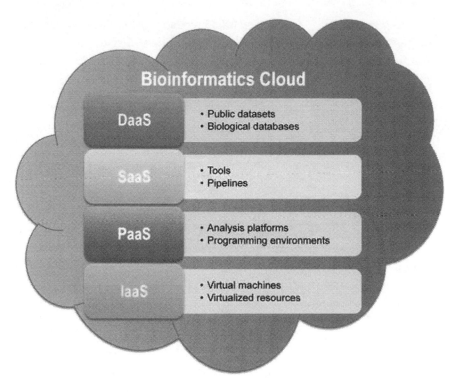

FIGURE 1.3   Bioinformatics cloud.

1.3 BIOINFORMATICS CLOUD COMPUTING SERVICES

There are a number of companies that offer their services in the fields of bioinformatics cloud analysis. Each of them offers a wide range of products to facilitate the analysis, and some have made a widely known name for themselves in the bioinformatics industry. A few of these companies and the solutions they offer are discussed below in detail.

1.3.1 Data as a Service (DaaS)

Data as a Service (DaaS) is one of the latest kinds of services currently under investigation in the cloud computing community. It aims to overcome the limitations of existing data-technology approaches in which the location of the stored and accessed data crucial for sharing and processing [5]. DaaS allows the processing and sharing of data with an abstract approach, and hence the exact location of the repositories is not relevant. Some of the bio-informatics approaches available in the DaaS market are:

1.3.1.1 EMBOSS

EMBOSS serves the microbiology and bioinformatics community, offering free access to a wide variety of free and efficient software tools for sequence analysis. It is an open-source

TABLE 1.1   EMBOSS Features

Name	EMBOSS
Platform	Desktop-based
Alternative names	European Molecular Biology Open Software Suite, EMBOSS explorer, Jemboss, wEMBOSS
Software type	Toolkit/suite
Interface	Graphical user interface
Operating system	Unix/Linux, Mac OS, Windows
Version	6.6.0
Restrictions to use	None
License	GNU General Public License version 3.0
Computer skills	Medium
Stability	Stable
Maintained	Yes

tool and permits the creation and release of new software in the open-source spirit. It is useful for sequence analysis and offers a wide variety of sub-tools, such as Labview, needle, transeq, backtranseq and many more, which modifies sequence analysis into a seamless whole (Table 1.1).

1.3.1.2 J-GLOBAL

J-GLOBAL aims to create links between science and technology information, and provides opportunities to make unexpected discoveries and gain knowledge. It includes tools to search synonyms and similar variable expressions to enlarge the field of users' query, presenting the results via a cloud-based delivery system (Table 1.2.).

TABLE 1.2   J-GLOBAL Features

Name	J-GLOBAL
Platform	Web-based
Computer skills	Basics
Interface	Web-user interface
Stability	Stable
Restrictions to use	None
Maintained	Yes

TABLE 1.3   Gene Set Builder Features

Name	Gene Set Builder
Platform	Web-based
Alternative names	GSB
Software type	Application/script
Interface	Web–user interface, application programming interface
Output data	A list of Entrez Gene, Ensemble, RefSeq, UniProt, or GeneLynx identifiers or a table populated with gene identifiers and descriptions
Output format	FASTA
Programming languages	Perl
Database management system	MySQL
Restrictions to use	None
Computer skills	Advanced
Stability	Stable
Registration required	Yes
Maintained	Yes

1.3.1.3 Gene Set Builder

Gene Set Builder (GSB) is a cloud-based web platform that manages and shares sets of open genes. GSB provides utilities to ease the workload (Table 1.3.). Its applications are:

- Annotating a pre-existing set of genes with a confident rating system.

- Sharing the newfound results with a defined community.

- Importing and exporting databases from a local computer in various specified formats.

1.3.1.4 BioDWH

BioDWH provides the users with ready-to-use parses that allow the extraction of data from publicly available life-science data sources, and stores the results in a data warehouse. It has platform and database independence, high usability, up-to-date integrated knowledge, and customization, the sole reason why it stands out from various other parsing software. BioDWH has already been used in the context of medical bioinformatics to integrate biomedical information with clinical data on rare metabolic diseases (Table 1.4.).

1.3.1.5 AWS Public Datasets

Amazon Web Services (AWS) offers a wide range of publicly available datasets that can be seamlessly integrated with the Amazon Web services to implement bioinformatics analysis. These datasets are delivered using a cloud-based web integrated platform (Table 1.5.).

TABLE 1.4   BioDWH Features

Name	BioDWH
Platform	Desktop-based
Interface	Command-line interface
Operating system	Unix/Linux, Mac OS, Windows
Software type	Package/module
Version	0.2
License	GNU General Public License version 2.0
Programming Language	Java
Restrictions to use	None
Computer skills	Advanced
Maintained	Yes
Stability	Stable

TABLE 1.5   AWS Public Dataset Features

Name	AWS Public Datasets
Platform	Desktop-based
Interface	Web-user interface
Parallelization	MapReduce
Computer skills	Basic
Restrictions to use	None
High-Performance Computing	Yes
Stability	Stable
Maintained	Yes

1.3.1.6 SeqHound

SeqHound provides the user with a locally hosted environment for bioinformatics research. It is a 3D structure database system that offers integrated biological sequences along with taxonomy and annotations. It also has the capability to retrieve specialized subsets of sequences, structures, and structural domains. It was built based on the data models and programming tools of the National Center for Biotechnology Information (Table 1.6.).

1.3.2 Software as a Service (SaaS)

A Software as a Service (SaaS) delivery model offers centrally hosted licensed software on a subscription basis. The SaaS provider allows the user to access on-premises applications, which are often unaffordable for smaller organizations due to high purchase and maintenance costs. Organizations do not have to worry about hiring skilled workers for maintenance, nor about the complexities of the connections, as the code is comparatively more straightforward and easier to implement [13]. SaaS is a pay-as-you-go model, enabling users to reduce expenses by efficiently managing their time on the software. The following are some of the bioinformatics approaches available in the SaaS market.

TABLE 1.6 SeqHound Features

Name	SeqHound
Platform	Desktop-based
Software type	Application/script
Interface	Command-line interface
Version	2.5
Operating system	Unix/Linux, Windows
Programming languages	C
Restrictions to use	None
Computer skills	Advanced
Maintained	No
Stability	Stable

1.3.2.1 Lhasa Cloud

Lhasa Cloud uses a cloud-based SaaS delivery system to provide access to the Lhasa software library. The system relies on Amazon Web Services (AWS). It offers software tools such as Vitic, a chemical repository, an information management system coupled in a single software and Mirabilis, which helps with the computation of purge factors of potentially mutagenic impurities of genes (Table 1.7.).

1.3.2.2 eCEO

eCEO (efficient Cloud-based Epistasis computing) is a cloud-based delivery model that is designed to identify epistatic interactions of single nucleotide polymorphisms (SNPs). The software is based on Google's MapReduce and Hadoop. This model has a practical advantage over other software as it allows the users to exhaustively search two-locus and three-locus epistatic interactions. The model is designed to retrieve top k-most interactions and leverages cloud computing with almost unlimited elastic computing resources (Table 1.8.).

TABLE 1.7 Lhasa Cloud Features

Name	Lhasa Cloud
Platform	Server-based
Software type	Application/script
Interface	Graphical user interface
Registration required	Yes
Operating system	Unix/Linux
License	Commercial
Restrictions to use	License purchase required
Computer skills	Medium
Stability	Stable
Maintained	Yes

TABLE 1.8   eCEO Features

Name	eCEO
Alternative names	efficient Cloud-based Epistasis computing
Platform	Desktop-based
Interface	Command-line interface
Software type	Application/script
Parallelization	MapReduce
Restrictions to use	None
Operating system	Unix/Linux
Computer skills	Advanced
High-performance computing	Yes
Stability	No
Maintained	No

1.3.2.3 StormSeq

StormSeq offers a cloud computing solution for informatics that performs well without any requirements for a parallel computing environment or extensive technical experience. StormSeq offers an open-source and open-access interface on the AmazonEC2 and can perform services like read cleaning, read mapping, annotation, and variant calling (Table 1.9.).

1.3.2.4 Crossbow

Crossbow is a cloud computing solution that is designed to search for single nucleotide polymorphisms (SNPs). It combines the accuracy of the SNP caller SOAP snp with the speed of the short-read aligner Bowtie to deliver a Hadoop-based tool equipped to perform alignment and SNP detection for multiple whole-human datasets. The software is known to deliver an accuracy of 98.9% on simulated individual chromosomes datasets, and a

TABLE 1.9   StormSeq Features

Name	StormSeq
Alternative names	Scalable Tools for Open-Source Read Mapping
Platform	Desktop-based
Interface	Command-line interface
Software type	Application/script
Parallelization	MapReduce
Operating system	Unix/Linux
Restrictions to use	None
Computer skills	Advanced
High-performance computing	Yes
Stability	No
Maintained	No

TABLE 1.10   Crossbow Features

Name	Crossbow
Platform	Desktop-based
Software type	Framework/library, pipeline/workflow
Interface	Command-line interface
Parallelization	MapReduce
Operating system	Unix/Linux
License	GNU General Public License version 3.0
Restrictions to use	None
Computer skills	Advanced
Version	1.2.1
Requirements	Bowtie, SOAPsnp
High-performance computing	Yes
Stability	Stable
Maintained	Yes

concordance better than 99.8% with the Illumina 1 M BeadChip assay of a sequenced individual (Table 1.10.).

1.3.2.5 CloudBurst
CloudBurst is a program that maps next-generation single-end sequence data to reference genomes and reports all alignments for each read up to a user-specified number of differences, including both indels and mismatches. It enables users to filter the alignments in order to identify the single best non-ambiguous alignment for each read (Table 1.11.).

TABLE 1.11   CloudBurst Features

Name	CloudBurst
Platform	Desktop-based
Software type	Framework/library
Interface	Command-line interface
Version	1.1.0
Operating system	Unix/Linux
Parallelization	MapReduce
Programming languages	Java
Restrictions to use	None
License	Artistic License version 2.0
Computer skills	Advanced
Maintained	Yes
High-performance computing	Yes
Stability	Stable

1.3.3 Platform as a Service (PaaS)

A Platform as a Service (PaaS) delivery model offers a web-based online platform that allows developers to develop, build, run, and manage various services. The PaaS provider provides the user with the hardware and software tools necessary to build applications [11]. The PaaS provider hosts the software and hardware on its own infrastructures and only provides the developers with the services. These services can be accessed over the internet with the help of a web browser. PaaS is available as both a pay-as-you-go model and a one-time-pay model [14]. PaaS is economical and suitable for small organizations, as it allows users to reduce the investment costs to zero. The following are among the variety of PaaS approaches in the bioinformatics market.

1.3.3.1 DNAnexus

DNAnexus allows next-generation sequence (NGS) analysis as well as NGS visualization, and manages genomic data. It delivers a system that can be used to create various projects, run apps and analysis, build workflows, and share them with other users at various access levels (Table 1.12.).

1.3.3.2 Magallanes

Magallanes is a platform-independent Java library that allows the user to discover various bioinformatics web services along with their associated data types with the aid of its pre-defined algorithms. It enables the connectivity of available and compatible web services into workflows and is equipped to sequentially process data so as to obtain a desired output for the given input. It is available both as an API-based application and with a graphical user interface (Table 1.13).

1.3.3.3 Google Genomics

Google Genomics is a platform that allows users in the scientific community to organize, share, and access the world's available genomics information. It employs the same technologies that are used by Google's search and map applications in order to maintain the security of the processes, data, and complex databases. It also facilitates easy sharing and allows the

TABLE 1.12   DNAnexus Features

Name	DNAnexus
Platform	Desktop-based
Interface	Command-line interface
Software type	Framework/library
Parallelization	MapReduce
Operating system	Unix/Linux
Computer skills	Advanced
Restrictions to use	License purchase required
High-performance computing	Yes
Stability	Stable
Maintained	Yes

TABLE 1.13 Magallanes Features

Name	Magallanes
Platform	Desktop-based
Alternative names	Multi-Architecture Resources Discovering
Interface	Graphical user interface, application programming interface
Software type	Package/module
Operating system	Unix/Linux, Windows
Restrictions to use	None
Programming languages	Java
Maintained	Yes
Computer skills	Medium
Stability	Stable

TABLE 1.14 Google Genomics Features

Name	Google Genomics
Platform	Desktop-based
Interface	Command-line interface
Software type	Package/module
Computer skills	Advanced
Restrictions to use	License purchase required
Stability	Stable

user to run multiple searches and experiments in parallel. The tool supports open industry standards specified by the Global Alliance for Genomics and Health (Table 1.14.).

1.3.3.4 Syapse

Syapse offers a platform dedicated to information management in the field of oncology. Syapse Oncology can be used not only by researchers for analysis but also by clinical professionals, as it offers an interface to communicate with patients. It compiles a set of features to harmonize and share molecular as well as clinical data and workflows, enabling researchers to investigate new treatments and to determine possible trials (Table 1.15.).

1.3.3.5 BioServices

Bioservices is a Python-based library that enables the user to access major bioinformatics web services. It provides access to about 20 web services that can be used individually or in combination, enabling users to complement external and foster the development of new workflows. The web services offered include UniProt, ChEMBL, KEGG, Ensemble, MUSCLE, BioModels, NCBIBlast, and WikiPathway (Table 1.16.).

TABLE 1.15   Syapse Features

Name	Syapse
Platform	Desktop-based
Interface	Graphical user interface
Software type	Framework/library
Computer skills	Medium
Parallelization	MapReduce
Operating system	Unix/Linux
Restrictions to use	License purchase required
High-performance computing	Yes
Stability	Stable
Registration required	Yes
Maintained	Yes

TABLE 1.16   BioServices Features

Name	BioServices
Platform	Desktop-based
Software type	Framework/library
Interface	Command-line interface
Version	1.4.17
Operating system	Unix/Linux
Programming languages	Python
Restrictions to use	None
License	BSD 3-clause "New" or "Revised" License
Computer skills	Advanced
Maintained	Yes
Stability	Stable

1.3.3.6 Eoulsan

Eoulsanis a framework that aims to ease high-throughput sequencing (HTS) data analysis with distributed computation. The application includes batch analyses as well as a full automation process that is equipped to manage both external file locations and the distributed file system. It is available in three runnable modes: local cluster, standalone, or cloud computing on Amazon Elastic MapReduce (Table 1.17.).

1.3.4 Infrastructure as a Service (IaaS)

An Infrastructure as a Service (IaaS) delivery model offers virtualized applications and computer resources over the cloud-based platform. The IaaS provider provides the user with high-level APIs that help them access various physical computing resources according to their requirements [11]. The user can choose various components of the system, such as the configuration of the CPU, memory, or storage of the computer, and can alter these

TABLE 1.17 Eoulsan Features

Name	Eoulsan
Platform	Desktop-based
Software type	Framework/library
Parallelization	MapReduce
Interface	Command-line interface
Operating system	Unix/Linux
Computer skills	Advanced
Programming languages	Java
Restrictions to use	None
License	GNU Lesser General Public License version 3.0
High-performance computing	Yes
Stability	Stable
Maintained	Yes

configurations easily using their interface to generate a simple request to the cloud server. IaaS is economical and avoids the complexities and expenses of managing devices, servers, and other hardware and infrastructure [11]. The following are among the bioinformatics approaches available in the IaaS market.

1.3.4.1 Google Compute Engine
Google Compute Engine (GCE) is an IaaS delivery model that allows users to run multiple large-scale workloads on Google-hosted Linux virtual machines (Table 1.18.).

1.3.4.2 GoGrid
GoGrid is one of the leading IaaS providers and provide services "on the go." It offers a hybrid cloud, a private cloud, and dedicated infrastructure services that are designed to serve the complex needs of a user (Table 1.19.).

TABLE 1.18 Google Compute Engine Features

Name	Google Computation Engine
Platform	Desktop Based
Interface	Command-line interface
Software type	Framework/library
Parallelization	MapReduce
Operating system	Unix/Linux
Computer skills	Advanced
Restrictions to use	None
High-performance computing	Yes
Maintained	Yes
Stability	Stable

TABLE 1.19   GoGrid Features

Name	GoGrid
Platform	Desktop-based
Interface	Command-line interface
Software type	Framework/library
Parallelization	MapReduce
Operating system	Unix/Linux
Computer skills	Advanced
Restrictions to use	None
High-performance computing	Yes
Stability	Stable
Maintained	Yes

TABLE 1.20   HP Helion Features

Name	HP Helion
Platform	Desktop-based
Interface	Command-line interface
Software type	Framework/library
Parallelization	MapReduce
Operating system	Unix/Linux
Computer skills	Advanced
Restrictions to use	None
High-performance computing	Yes
Maintained	Yes
Stability	Stable

1.3.4.3 HP Helion

HP Helion delivers a developer-oriented, open-source-based, business-friendly infrastructure that is suitable to fulfill the analysis needs of a bioinformatics specialist (Table 1.20.).

1.3.4.4 Joyent

Joyent is the world's only container-native public cloud and allows its users to securely deploy and operate containers with bare-metal speed. The Joyent team pioneered hybrid as well as public cloud computing, grew and nurtured Node.js into a well-practiced standard for mobile, web, and IoT architectures, and was among the first to embrace and industrialize compute-centric object storage. In a development that will take the technological world by storm, Joyent is currently working on serverless computing (Table 1.21.).

TABLE 1.21   Joyent Features

Name	Joyent
Platform	Desktop-based
Interface	Command-line interface
Software type	Framework/library
Parallelization	MapReduce
Operating system	Unix/Linux
Computer skills	Advanced
Restrictions to use	None
High-performance computing	Yes
Stability	Stable
Maintained	Yes

1.4 CSIM ARCHITECTURE

Cloud Simulation (CSIM) is an architecture that builds a dynamic, flexible, robust, adaptable, and scalable system that is easy to maintain by combining SOA and intelligent agents with the cloud computing approach. CSIM is an intelligent, distributed system that allows cloud services to communicate with each other without any time or location restrictions [15]. Communication is even possible through mobile devices. The CSIM architecture does not integrate the functionalities of the systems with its agents: they do not act as controllers or coordinators and have no right to invoke services. The agents merely increase the reasoning capabilities of the system and allow it to handle the cloud-based services in the context of the characteristics of the system, which are capable of changing dynamically over time [15].

CSIM agents are autonomous, have reasoning and reacting capabilities, and possess pro-activity, mobility, and organization. All of these characteristics allow them to cover various needs of dynamic systems such as adaptable interfaces and ubiquitous communication and computing. CSIM combines intelligent agents with a cloud computing approach built on top of web services to create an innovative architecture facilitating high levels of human–system–environment interaction [15]. It also provides customization and advanced flexibility to easily modify, add, or remove services on demand. The prime objective of CSIM is to deliver a flexible system that has a higher ability to recover from errors and allows behavior to be changed during the time of execution.

The 4 basic blocks of the CSIM architecture are:

- *PaaS (Platform as a Service)* consists of all the custom applications that can be used to take advantage of the system functionalities. PaaS applications are dynamic and adaptable, reacting according to the situation. They can also be accessed on remote mobile devices with less processing capability, in contrast to other locally accessible applications [15]. This is possible because the computational tasks are largely delegated to the agents and other services associated with the systems services.

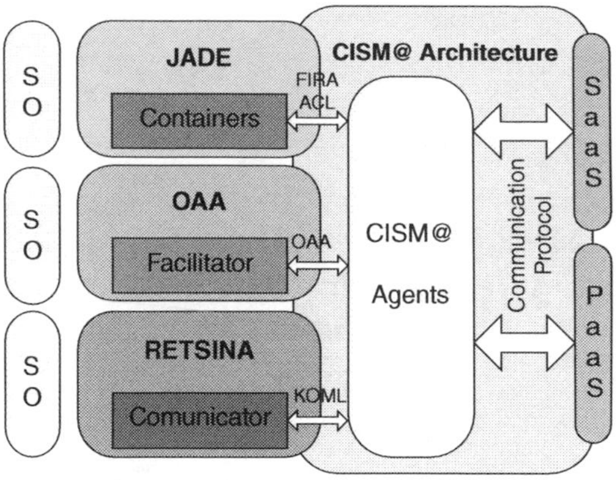

FIGURE 1.4   CSIM architecture.

- *Agent Platform* is the core of the CSIM architecture. It can be referred to as the set of agents that contains the agents that are predefined by CSIM itself. It also deals with virtual organization to aid massive data analysis. The virtual organization of the agents is established in the case study function, generating laboratory personnel-like behavior for microarray data analysis of an organization.

- *SaaS (Software as a Service)* represents the activities that the CSIM architecture has to offer. These services are specifically designed to be invoked locally as well as remotely and can be organized as web services, local services, cloud services, or individual standalone services [15]. Each service can invoke or make use of other services to provide the required functionalities. CSIM has a flexible and scalable directory of services that can be invoked, modified, added, or eliminated dynamically as per user requirements. The agent platform itself includes case-specific services.

- *Communication Protocol* establishes direct communication between applications and services and the platform agent. The protocol is open and language independent, which facilitates its ubiquitous communication capabilities [15]. This protocol is based on SOAP specification and captures all messages exchanged between the platform and the services and applications. All external communications also follow the same protocol . Communication among agents in the platform follows the FIPA Agent Communication Language (ACL) specification. The CSIM cloud computing architecture is illustrated in Figure 1.4.

1.5 IMPLEMENTATION CHALLENGES

The main challenges facing cloud-based implementation of bioinformatics are security and privacy, as biological data is not available publicly, and accessing patients' medical records without their consent is regarded as a breach of trust. A variety of legal and economic checks are implemented to ensure privacy and security of data, which further limits data accessibility, and at times increases the costs of implementation [5]. This section discusses these implementation challenges in detail.

1.5.1 Security and Privacy

Security and privacy of the individual are a must, and accessing their medical records for scientific gain represents a breach of their privacy and trust. There are eight aspects of security—integrity, confidentiality, authenticity, accountability, audit, non-repudiation, anonymity, and unlikability—and it is important that available data-sets meet the standards required for all these aspects to protect a person's right to privacy [5].

- *Integrity*: Data integrity ensures the consistency and accuracy of the data over its entire life-cycle and is of concern to every organization that deals with storing, accessing, managing, and processing data. Since bioinformatics deals with large chunks of data, integrity is one of the main concerns for cloud implementation of bioinformatics.

- *Confidentiality*: Confidentiality only allows authorized and registered personnel to access and modify data. It is a multi-fold process, with checks implemented on various security levels based on the structural needs of the organization. Protected personal clouds can be useful when talking about confidentiality in bioinformatics clouds.

- *Authenticity*: Data authentication itself is a two-fold process that deals with collecting information about the origins of data and then validating its integrity. It is mainly based on communication and can be identified as an implementation check for integrity.

- *Accountability*: Data accountability ensures that the organizations involved own responsibility for the data they access. Organizations are required to protect the privacy of users, and it is their responsibility to ensure that no misuse takes place at their end.

- *Audit*: Data audits are conducted to assess how the data is being used by the company, and whether the company's data is fit for a given purpose. This is achieved by profiling data and then assessing the impacts of incomplete, partial, or incorrect data on the company's performance and profits to identify the extent and amount of data required.

- *Non-repudiation*: Non-repudiation ensures that no involved party is able to deny their agreement/involvement in a contract or communication by denying the authentication of their signature. Parties involved are to own full responsibility for messages originated by them.

- *Anonymity*: Data anonymity maintains the privacy of the user to a great extent. It involves encryption or removal of personally identifiable information of users participating in biological experiments, and enables organizations to access their medical/biological records with no possibility of personal identification.

- *Unlikability*: This aspect of security limits the ability of the organization to establish links between two data-sets or multiple data points of the same data-set. This protects complete user information and works to their advantage.

1.5.2 Legal Aspects

Various countries have laws that regulate and protect the security of patients by protecting their medical records and ensuring that nobody other than doctors or the patient themselves have free access to these records [1]. Countries such as the US and Canada have strict laws regulating the privacy of medical records, but countries like India still lag behind, making patient medical data vulnerable to exploitation by anyone and everyone. The US Health Insurance Portability and Accountability Act (HIPAA) limits companies from disclosing personal health data to third parties, while the Canadian Personal Information Protection and Electronic Documents Act (PIPEDA) prohibits organizations from collecting, using, or disclose personal information in commercial activities [1].

In India, section 43(a) and section 72 of the Information Technology Act provide the basic framework and rules for the protection of personal information, including medical records [1]. The Act establishes various other rules concerning sensitive personal information to be followed by any entity that collects, stores, or deals with sensitive information, such as passwords, health conditions, medical records, financial information, sexual orientation, and biometric records.

The Act mandates organizations ("bodies corporate") to undertake reasonable procedures to protect sensitive personal data or information, and section 72 protects personal information from unlawful disclosure in a breach of contract. The "body corporate" is defined as "a firm, sole proprietorship or other association of individuals engaged in commercial or professional activities." A large number of people in India seek medical attention from public hospitals as not everybody can afford private healthcare [1]. Public hospitals and NGOs in India are not equipped to handle individuals' privacy concerns, hence a breach of security is very likely.

REFERENCES

[1] S. Sharma, K. Kaur, and A. Singh. Role of cloud computing in bioinformatics. *International Journal of Computer Techniques*, 3(3):1–4, May–June 2016.

[2] D. Field, B. Tiwari, T. Booth, S. Houten, D. Swan, N. Bertrand, and M. Thurston. Open software for biologists: from famine to feast. *Nature Biotechnology*, 24(7):801–804, 2006.

[3] E. Afgan, D. Baker, N. Coraor, B. Chapman, A. Nekrutenko, and J. Taylor. Galaxy cloudman: delivering cloud compute clusters. *BMC Bioinformatics*, 11(Suppl 12):S4, 2010.

[4] J. Goecks, A. Nekrutenko, J. Taylor et al. Galaxy: a comprehensive approach for supporting accessible, reproducible, and transparent computational research in the life sciences. *Genome Biology*, 11(8):R86, 2010.

[5] K. Krampis, T. Booth, B. Chapman, B. Tiwari, M. Bicak, D. Field, and K. E. Nelson. Cloud bio-linux: pre-configured and on-demand bioinformatics computing for the genomics community. *BMC Bioinformatics*, 13(1):42, 2012.

[6] B. Langmead, M. C. Schatz, J. Lin, M. Pop, and S. L. Salzberg. Searching for snps with cloud computing. *Genome Biology*, 10(11):R134, 2009.

[7] P. Lawrence, Ed., *Workflow Handbook*. John Wiley & Sons, Inc., New York, NY, 1997.

[8] B. Liu, R. K. Madduri, B. Sotomayor, K. Chard, L. Lacinski, U. J. Dave, J. Li, C. Liu, and I. T. Foster. Cloud-based bioinformatics workflow platform for large-scale next-generation sequencing analyses. *Journal of Biomedical Informatics*, 49, 2014.

[9] A. Matsunaga, M. Tsugawa, and J. Fortes. Combining MapReduce and virtualization on distributed resources for bioinformatics applications. In *eScience, 2008. eScience'08. IEEE Fourth International Conference on IEEE*, Indianapolis, Indiana, 222–229, 2008.

[10] H. Nordberg, K. Bhatia, K. Wang, and Z. Wang. Biopig: a Hadoop-based analytic toolkit for large-scale sequence data. *Bioinformatics*, 29(23):3014–3019, 2013.

[11] M. C. Schatz. Cloudburst: Highly sensitive read mapping with MapReduce. *Bioinformatics*, 25(11):1363 1369, 2009.

[12] A. Schumacher, L. Pireddu, M. Niemenmaa, A. Kallio, E. Korpelainen, G. Zanetti, and K. Heljanko. Seqpig: simple and scalable scripting for large sequencing datasets in Hadoop. *Bioinformatics*, 30(1):119–120, 2014.

[13] M. S. Wiewiórka, A. Messina, A. Pacholewska, S. Maffioletti, P. Gawrysiak, and M. J. Okoniewski. Sparkseq: a fast, scalable, cloud-ready tool for the interactive genomic data analysis with nucleotide precision. *Bioinformatics*, 30(18):343, 2014.

[14] S. Zhao, K. Prenger, L. Smith, T. Messina, H. Fan, E. Jaeger, and S. Stephens. Rainbow: a tool for large-scale whole-genome sequencing data analysis using cloud computing. *BMC Genomics*, 14(1):425, 2013.

[15] J. Bajo, C. Zato, F. de la Prieta, A. de Luis, and D. Tapia. Cloud computing in bioinformatics. *Advances in Intelligent and Soft Computing*, 79:147–155. doi:10.1007/978-3-642-14883-5_19.

Intelligent Cloud Computing and Bioinformatics Data Analysis

Deepak Kumar Sharma, Prastuti Upadhaya, and Satwik Bhardwaj

CONTENTS

2.1 INTRODUCTION

This section introduces intelligent cloud computing (ICC) and bioinformatics data analysis, and explores reasons for the application of "intelligent" clouds to data analysis.

2.1.1 Introduction to Intelligent Cloud Computing

The advent of cloud computing dramatically altered the conventional computing landscape by providing a level of abstraction or separation between computing resources and their associated technical framework [1], enabling demand-compatible self-service, location-independent resource management, ubiquitous network access, rapid elasticity, pay-per-use facility and requirement-dependent quality of service. Since the first cloud computing infrastructure, Amazon Web Service (AWS), was introduced in 2006 by Amazon, demand has risen, with Microsoft's implementation of .NET technology on Azure, the provision of Cloud Foundry by VMware and Google's development of the AppEngine. Growing dependence on this technology and the increasing number of Web 2.0 applications have led to the transfer of voluminous amounts of data over the cloud. Moreover, accountability and the responsibility for maintaining the computing infrastructure have shifted from consumers to service providers. The emerging challenge of an increasingly sophisticated cloud computing infrastructure has led the research community to seek ways of making this technology "intelligent".

Intelligent cloud computing (ICC) may be defined as cloud computing technology that is built to achieve enhanced functionality through artificial intelligence. It makes use of the cloud's inherent characteristics, with features like latency optimization and automated resource allocation offered by various distributed software agents. This advance offers great promise for both researchers and clients, particularly those involved in the field of informatics and data analysis. This chapter explores various approaches to ICC, its architecture, scope, trends and challenges, and its application in the field of bioinformatics data analysis.

2.1.2 Contemporary Advancements in Bioinformatics Data Analysis

Bioinformatics research and its applications include analysis of the nuances of molecular sequencing and data related to genomics; gene prediction; molecular folding, modelling and design; database development; building biological networks and data management systems; development of analysis tools and software; mining of biomedical data records and text; and education in bioinformatics.

In 1974, DNA sequences began to be submitted to GenBank [2]. However, the more significant development in the storage of DNA sequences in databases was the inclusion of web-based search algorithms that allowed comparison of target DNA sequences. The resulting computer software, GENEINFO [2], allowed rapid searches for sequences indexed in a database, enabling them to be matched with the sequence in question. There has been an enormous increase in the amount of genetic data stored at various institutes across the world (see example in Figure 2.1).

Software and analysis tools are integral to the study of bioinformatics data. Since the 1980s, there has been substantial development in the software, starting from simple

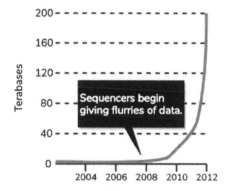

DATA EXPLOSION
The amount of genetic sequencing data stored at
the European Bioinformatics Institute takes less
than a year to double in size.

FIGURE 2.1   Increase in the amount of gene sequencing data stored over the years at the European Bioinformatics Institute.

command-line tools to now, much more complex graphical program packages and other forms of web services [3, 4]. Researchers have stressed the importance of analyzing super-datasets that require very large storage capacity and faster analysis than that provided by traditional databases. Cloud computing is increasingly used in bioinformatics data analysis, as it is more cost-effective than traditional research methods where large amounts of data are downloaded from public repositories and personal infrastructure is used for data analysis. Software and data alike are stored in the cloud, so that researchers do not have to download the software or be adept at the installation of operating systems. The use of cloud computing, with its lower funding levels, has led to more and more people being included in healthcare research [5].

2.1.3 Reasons for the Use of Intelligent Cloud Computing in Bioinformatics

Rapid advancements in biological and biomedical data acquisition, such as DNA sequencing, have led to a sharp decline in procedural costs. This, in turn, has led to huge genome datasets being generated in real time. The vast scale of bio information can be estimated from the fact that an uncompressed human genome occupies about 3 gigabytes [6], and the world population is increasing even as you read this text. The development of next-generation sequencing (NGS) technologies has made DNA sequencing cheaper than storage and computational costs. Studies show that since 2010, using DNA sequencing to acquire the data has become cheaper than the disk storage costs for the same data, and the gap is increasing exponentially [6].

This rapid growth in bio information, coupled with the need for analysis, has compelled the research community to look for storage facilities which are larger than traditional databases such as GenBank at the National Center for Biotechnology Information (NCBI) [7], the microarray database Array Express [8], the European Bioinformatics Institute EMBL database [9], the Gene Expression Omnibus (GEO) [10], DNA Data Bank of Japan (DDBJ) [11] and the Sequence Read Archive (SRA) [12]. Storage is not the only concern regarding

TABLE 2.1   Terminology and Abbreviations

Symbol	Definition
MAS	Multi-agent systems
QoS	Quality of service
AWS	Amazon Web Services
NGS	Next-generation sequencing
HPC	High-performance computing
MR	MapReduce
MST	Minimum spanning tree
SAM	Significance analysis of microarrays
AI	Artificial intelligence

efficient architectures for bioinformatics data analysis, however. Conventional data-retrieval applications such as Ensemble, the University of California at Santa Cruz (UCSC) Genome Browser [13] or Galaxy [14] also face computational challenges. Data storage interfaces, including archives, integrators and power users, are faced with fluctuating demands for data which can lead to inefficiencies, with systems designed to handle peak computational requirements at all times but exhausting them even when demand is low, leading to wastage of computational resource.

It is thus that the problems posed by the sophistication of bioinformatics data analysis and the need for an efficient architecture have led to the involvement of ICC in this field of science.

2.1.4 Terminology and Abbreviations

Table 2.1 defines the terms most used in this chapter.

2.2 PRINCIPAL DISCIPLINES OF INTELLIGENT CLOUD COMPUTING AND BIOINFORMATICS DATA ANALYSIS

This section analyzes the architecture of ICC systems and current prevalent frameworks for bioinformatics data analysis.

2.2.1 Intelligent Cloud Computing Architecture

An intelligent cloud computing system has become an indispensable resource in today's tech world. The need to optimize the processing, analysis, storage and sharing of big data has brought about the advent of artificial intelligence (AI) in cloud computing. As defined by AI experts, ICC is the maximum optimization of cloud computing resources with the help of "intelligent methods and technologies" [15–18]. Figure 2.2 provides a basic idea of an intelligent system and the relationship between it and the user. The user will provide the facts and get expertise accumulated by the system in return.

The three main components of an ICC system (see Figure 2.3) are [19]:

1. **Data Source:** The source of the data used in the analysis is not just external; data from within the corresponding cloud computing system are also included. These data are divided into different layers that assist management tactics optimization: the physical

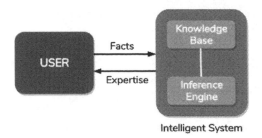

FIGURE 2.2 The basic idea of an intelligent system.

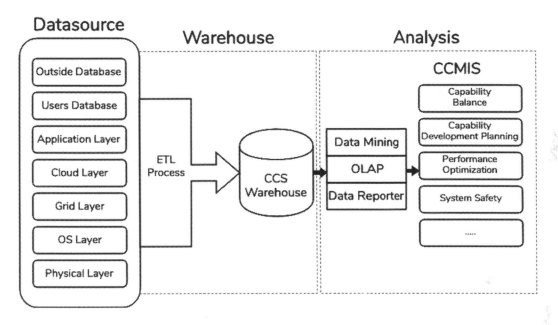

FIGURE 2.3 Components of an intelligent computing system.

layer, grid-middleware layer, operating system layer and cloud-middleware layer. Since there are too many factors affecting the data, collection and analysis of operational data is extremely important.

2. **Data Warehouse:** Data warehouses tend to be subject oriented, which implies that information collected from the sources has to be stored by relevance and be organized. The data are extracted, transformed and subsequently loaded in preparation for the next step, analysis.

3. **Analysis:** The cloud computing management information system (CCMIS) emphasizes a decision-making process that is nonstructural. Structured decisions may be reached without any manual intervention. The focus is largely on capability balance, resource planning, system safety and other issues of strategic relevance.

The goal of an AI system is to make programs that display intelligent behaviour once the demands of the intelligence required in that program are understood [15].

2.2.2 Current Frameworks in Bioinformatics Data Analysis

This section discusses the various frameworks in bioinformatics data analysis, the types of service models available in cloud computing, and when to use which one.

2.2.2.1 MapReduce with Hadoop for Big Data Analysis

MapReduce (MR) is a programming model devised for advanced high-performance computing (HPC) in the cloud, which is suitable for big data analysis in the field of bioinformatics. As the name suggests, this model has two stages [20]:

1. Map Stage: The input to MapReduce software is first split into smaller sub-inputs and then passed to a mapping function that produces an output data for each corresponding sub-input.

2. Reduce Stage: The outputs from the mapping function are consolidated and organized through a process known as shuffling. The shuffled data sets are then fed into a *reducer,* which integrates the sub-datasets to yield the final result.

Thus, the MR model proposes *parallel* processes to efficiently analyze data by scheduling and dispatching sub-tasks. As shown in Figure 2.4, each process handles only a subset of data rather than the entire data set, significantly reducing processing time. The following illustration explains the MR structure.

This approach to data analysis has been implemented by an open-source software framework, Hadoop. As MR requires high processing power and a large storage facility,

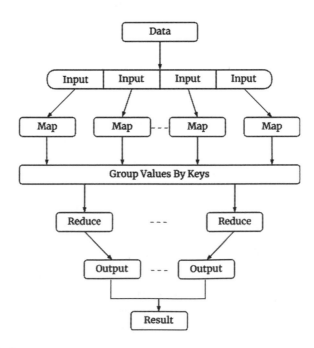

FIGURE 2.4   MapReduce structure.

Hadoop has taken advantage of cloud computing resources including Amazon Elastic Map/Reduce [21] and IBM Infosphere Big Insights [22]. In April 2019 Microsoft Azure also announced the release of Hadoop 3.0 on its Azure HDInsight analytics service [23]. With increasing support for Hadoop from leading cloud computing infrastructures, the transition from the conventional implementation of MapReduce to cloud-based platforms is imminent.

2.2.2.2 Applications of Service Models

Accessibility and connection are vital in the modern world; cloud computing aims to ensure that enterprises work to the requirements and limitations of the clients. Depending on the needs and outcomes, different types of service models are used. Figure 2.5 displays the different types of service models, which aspects have to be handled by the user, and what is handled by the providers. Cloud service models are mainly of three types [24]:

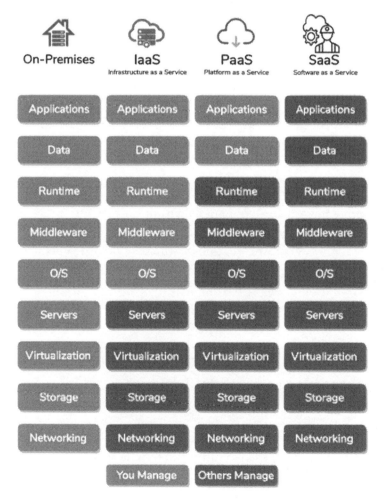

FIGURE 2.5   Different types of service models showing distribution of the management of resources.

- Software as a Service (SaaS) grants access to web servers based in the cloud. The entire computing stack is controlled by the vendor, which can be accessed via a search engine. The applications are maintained on the distant cloud and can be accessed on payment of a subscription, or are free for limited access. SaaS eliminates the need to install software and applications on every computer, and the vendor looks after the application, providing maintenance and the required support. Examples include Dropbox, Google G Suite and Microsoft Office 365 [25].

- Infrastructure as a Service (IaaS) makes provision for the entire infrastructure to be set up and customized in the cloud. A service provider that provides IaaS cloud services can supply a huge range of computing infrastructure, including storage, networking hardware, servers, support and maintenance.

  Enterprises can opt for computing resources based on their own requirements, without the need to set up any hardware at the workplace. Among the leading IaaS cloud service providers are Amazon Web Services, Google Compute and Microsoft Azure [25].

- Platform as a Service (PaaS) is a cloud service providing a virtual runtime environment where specific applications can be developed, organized and tested. The use of PaaS greatly simplifies software development for enterprises. The company or a platform provider can easily manage the cloud resources in the form of storage, servers and the underlying networking. Well-known examples include AWS Elastic Beanstalk and Google App Engine. PaaS is subscription based and comes with flexible pricing options, depending upon the requirements of the client [25].

2.2.2.3 Cloud Developments in Translational Biomedical Sciences

Bioinformatics generates a huge amount of data from various sources, such as gene expression, genotyping or NGS data. Raw sequence data is usually stored by researchers in the Sequence Read Archive (SRA), the volume of which stood at 1.6 petabytes in 2013 [26]. Data classified on the basis of genomic factors are used to identify genetic factors causing disease or impacting health. Data determined by diagnosis, on the other hand, come from insurance records, prescription data, healthcare information, and so on. The following are resources for the two different types of biomedical data that currently exist in the cloud:

- *Genomic-driven data*. Efficient solutions are needed in terms of storage, analysis, data transfer and computation. The BioVLab infrastructure [27, 28] available on the cloud, for example, uses a virtual collaborative laboratory that provides researchers with facilities that are remote, like computing or data storage, rather than being readily available. The MapReduce workflow is applied by the Crossbow genotyping program on Hadoop, and it is used to introduce Bowtie copies in parallel [29]. When the reads are produced, Hadoop begins the workflow, calls to sort and then aggregate the alignments.

- *Diagnosis-driven data*. Biomedical data tend to be heterogeneous and heavy; hence high-quality data processing becomes increasingly difficult. Computer-intensive challenges can be countered by the graphics processing unit (GPU), which has two main application programming interfaces (APIs): CUDA and OpenCL. The architecture includes a number of microprocessors, with stream processors. The work is divided into threads and blocks. Each multiprocessor has a number of threads, defined by the user, and the blocks run on multiprocessors. Stream processors have a single unit of fetch-decode inside the same multiprocessor, and hence threads are forced to execute in lockstep. To increase flexibility, a thread may split from the execution path.

Although cloud systems in bioinformatics are advancing every day, emerging concerns over data security and safety are becoming the main issue with this technology from a commercial perspective [30].

2.3 BIOINFORMATICS DATA ANALYSIS METHODOLOGIES COMPARED

In this section, we compare two bioinformatics data analysis methodologies using cloud computing – AiNET and microarray data analysis – and assess the potential transformation of the current data analysis paradigm.

2.3.1 AiNET (Artificial Immune NETwork)

AiNET is an artificial immune system, based on immune network theory, first proposed by Niels Jerne in 1974 [31, 32]. It compresses data using methods proposed by immune network theory and the clonal selection principle [33]. It reduces complexity by decreasing data cardinality, and filters the outliers, using various prototypes to represent the wide variety of data clusters. The procedure involves two main steps: AiNET adaptation; and the introduction of diversity and immune network interactions. The fundamental principle of immune network theory followed here is that immune cells have the capacity to recognize each other. Once an affinity between memory cells has been established, cells with an affinity more significant than a predetermined threshold have to be suppressed to reduce inconsistency in the network. The introduction of new cells into the population of immune cells creates diversity. The next step involves defining a mechanism that has the capability to detect separations inherently present in the spatial distribution of the antibodies involved [33].

The minimum spanning tree (MST), derived from fundamental graph theory, has proved to be a driving force in data-clustering development. First, a set of points known to represent valid data points are inter-linked to produce an MST. This is followed by the removal of inconsistent and redundant links, producing a disconnected graph, as shown in Figure 2.6. Each sub-graph obtained corresponds to a unique data cluster. We therefore get an MST that not only defines the cluster members but also estimates the number of data clusters involved.

AiNET requires a lot of data processing and mathematical tools to remove inconsistent edges, determine probabilities and create the minimum spanning tree, which assists in data clustering and sorting, hence expediting further analysis. This may involve software traditionally installed by the researchers and hence may cost more. However, cloud

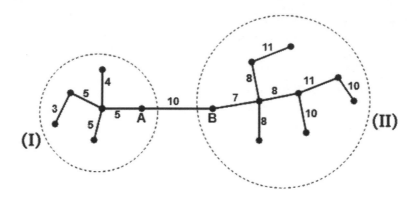

FIGURE 2.6   Minimum spanning tree for identifying data clusters.

computing provides the storage and software facilities for such applications together with solutions to complex function optimization problems, and cloud models help redesign the three major immune operators – cloning, mutation and suppression.

2.3.2 Microarray Data Analysis

The widespread application of gene expression profiling in the field of biotechnology has made microarray data analysis an essential technique for bioinformatics data analysis. It is a method of analyzing and interpreting data generated from DNA, RNA and, especially, microarrays. A microarray is a typical glass slide which is used to hold DNA molecules in an orderly fashion at specific positions or locations known as spots or features [34]. As shown in Figure 2.7(A), a spot is a site where about a million copies of the same DNA molecule corresponding to a unique gene are present. Technically, microarray data analysis is the concluding step in the processing of data generated by a microarray chip.

There are various ways that microarrays can be used to analyze gene expression, depending on the purpose and profile of the input dataset.

1. Comparison of sets of genes

   Microarrays may be used to compare gene sets on the basis of gene expression. This method detects significant differences in gene expression by taking size and variability into account. Gene expression is analyzed with the help of the final image obtained from the laser excitation of the hybridized cDNA labelling with dyes, as shown in Figure 2.7(B). It has widespread applications in the detection and profiling of genes in data labs.

2. Clustering

   Clustering – which may be hierarchical or non-hierarchical – is used to group genes on the basis of their expression or behavioural patterns. Hierarchical clustering may further be divided into agglomerative (proceeds by treating each object as a cluster) and divisive (proceeds by treating objects as discrete entities) clustering. Non-hierarchical clustering includes methods like K-means clustering (Figure 2.8), which employs algorithms to group genes into K groups.

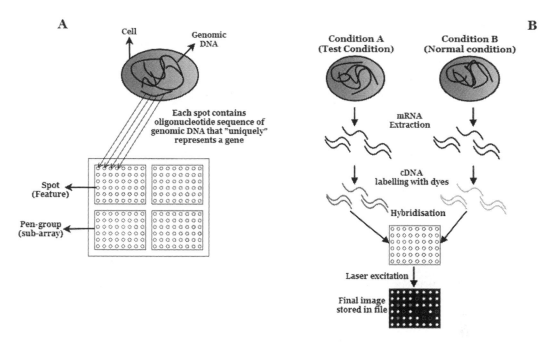

FIGURE 2.7   (A) A microarray containing a large number of spots from DNA. (B) Differential expression of genes.

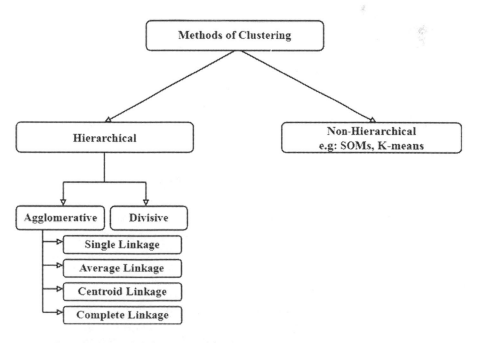

FIGURE 2.8   Classification of clustering methods.

The multitude of data generated by these techniques has led to the development of techniques like significance analysis of microarrays (SAM) for assessing the statistical significance of the change in gene expression. It is now possible to obtain the expression of millions of genes in a single experiment by using microarrays. With high levels of data involved, SAM helps determine whether the data generated is significant or not.

Computational tools for microarray data analysis are provided both by microarray manufacturers, such as Affymetrix and Agilent, and by open-source platforms such as BioConductor, which uses the R programming language. Although numerous tools for microarray data analysis have emerged in the last two decades, and techniques like SAM help filter out irrelevant data generated in DNA analysis, microarray data analysis lacks the high-performance computing and large data storage required, limiting further developments in bioinformatics data analysis.

2.4 CHALLENGES AND OPPORTUNITIES

This section explains the various challenges and opportunities faced by ICC and its application to the field of bioinformatics.

2.4.1 Technical Challenges and Scope for Improvement

Although the future of bioinformatics data analysis relies heavily on intelligent cloud computing, this paradigm shift poses a significant, if not overwhelming, number of challenges. Apart from the economic concerns raised by some technical reports [35], this transfer of platform also involves some technical challenges.

2.4.1.1 Workload Factoring

Growing dependence on cloud computing infrastructure in various fields will eventually lead to an exponential increase in the workload of servers. Although this issue is satisfactorily addressed by multiple instances of the same server running, even in different geographical locations, workload *factoring* still poses a technical threat. Incoming traffic to a particular domain will be directed to any of the instances of the server, but the factoring, or distribution, of the workload is not so simple. With an increase in the sophistication of servers and their functionalities, this challenge needs to be addressed with the utmost urgency.

2.4.1.2 Network Bandwidth

Network bandwidth is the biggest obstacle for genomics moving to the cloud. Limited bandwidth will be incompatible with the speed with which data is acquired using next-generation sequencing techniques. Consider a typical research institute that has a network bandwidth of around 1 gigabyte per second. On a given day, this bandwidth will be able to support the transfer of data over the internet at a speed of about 10 to 15 megabytes per second. Transferring a file of about 100 gigabytes of data acquired through enhanced DNA sequencing techniques to the database over the web will take at least a week. With the 10 to 15 gigabytes per second network bandwidth typically available in large universities and research institutions it will take less than a day but at the cost of blocking the entire bandwidth of the institution [6]. Thus, there is a significant mismatch between the desired and the actual transfer rates, which may lead to reluctance to move to the cloud on the part of an already established research institution.

2.4.1.3 Heterogeneous Distributed Database System

The spread of cloud computing across services has brought computing from varied fields to a common platform, leading to a significant rise in the interlinking of related databases, improving user experience and facilitating research. The complexity of this integrated environment containing heterogeneous databases is further aggravated by different operating systems, making exchange of information across the network intractable, if not impossible.

The research community is continuously looking for ways to overcome these technical challenges by making cloud computing "intelligent". Workload factoring is a really promising way forward in which *trespassing* workload, which refers to suddenly increased demands from the server at a particular time, is separated from *base* workload, the uniform and continuous workload experienced by the server at all times [36]. In order to expand in the field of genomics, cloud computing must offer flexibility in the flow of large data sets to overcome the challenge of network bandwidth. Here, cloud computing services can follow the model of the Protein Database [37], which used to accept atomic structure datasets on tape and floppy disk. The integration of non-uniform databases and servers requires detailed and researched solutions, one of which could be a protocol for regulating interactions in such distributed networks.

With ICC applications in bioinformatics data analysis continuing to evolve, there is great scope for improvement in this field.

2.4.2 Implementations and Applications

A series of waves in the development of the cloud can be identified. The first wave centred around PaaS and was provided by Google App Engine, Azure, and Heroku. The second wave was about IaaS, which enabled clients to acquire their own virtual machines and data storage. The next wave revolved around all-around data. With everything from big data to relational databases and graph databases, providers of cloud services supplied data platform services. Artificial intelligence has proved to be the next big development in the public cloud.

Cognitive computing is being used for APIs such as visual analytics, speech and text analysis, translation and detection, which are accessible via REST API endpoints to authorized developers. Cloud vendors are now even working with customized data for cognitive computing, which allows customers to supply their data to develop cognitive services and deliver specialized services. AI services and AI tools are becoming increasingly common among cloud service providers. AI in public cloud platforms is still underdeveloped, but it is playing a significant role in facilitating the adoption of data and computing services [38]. Figure 2.9 illustrates the different parts of an AI stack in a public cloud, showing the services, tools and infrastructure provided by AI.

Databases are being turned into vital "intel", with past buying patterns used to predict the behaviour of consumers. AI is not only a source of innovation but also a way to accelerate change.

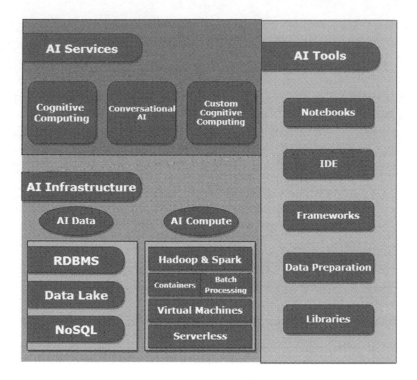

FIGURE 2.9   Public cloud AI stack.

2.5 FURTHER ADVANCEMENTS

The ever-evolving nature of intelligent cloud computing and the growing demand for high-performance computing opens up many opportunities for further advancements in the (largely unexplored) application of ICC to bioinformatics data analysis.

Cloud computing has witnessed dramatic evolution from the entire operating system (OS) holding virtual machines, used by VMware, to "capsules" called containers which wrap complete software, from heavy servers to serverless hosting, introduced by AWS in 2014, and from servicing monolithic applications to micro-servicing small "modules". These advancements have been complemented by DevOps, which combines developers and operations management for continuous integration and delivery. Cloud computing can also take advantage of multi-agent systems (MASs), with each technology complementing the other. There are two approaches to this idea. One uses these MASs for cloud implementation, improving service composition, service-level management and service-level agreement (SLA) negotiation due to their inherently distributed nature. The other approach concerns cloud-assisted software agent systems that can offload computing-intensive workloads to a high-performance server while maintaining a distributed software service [1]. Despite its apparent advantages, this concept of a shared domain is only followed by a few research activities [39–42].

Intelligent cloud computing has witnessed its latest development in the form of Amazon Macie [43], launched by AWS. This is a security service that uses machine learning to discover, classify and protect sensitive data in AWS. Macie paves the way for AI to be used to set up better infrastructure for the widespread application of cloud computing services.

One possible motivation for further advancements could be the sharing of computing infrastructure to address availability of computational resources and expertise in data management, such as the nationwide digital infrastructure designed to support the UK microbiology community. This shared electronic infrastructure caters to the needs of microbiologists across the country, providing an environment for the exchange and analysis of data through the Cloud Infrastructure for Microbial Bioinformatics (CLIMB) facility [44] . This approach is helping to maintain the pace of biotechnological advancements by tackling the explosion of available genome datasets for diagnostics and therapeutics [45]. A more recent development is BioVLAB [46], a bioinformatics computing architecture using Amazon Web Services and the Linked Environments for Atmosphere Discovery (LEAD)/ Open Grid Computing Environments (OGCE) scientific workflow system.

Further advancements in ICC should go hand in hand with its application in bioinformatics data analysis to meet the continuously evolving requirements and expectations of the microbiology community.

2.6 CONCLUSION

The potential of intelligent cloud computing for bioinformatics data analysis, in comparison with conventional approaches like microarray analytics, is widely accepted in the microbiology community. But this paradigm shift in the analysis framework requires smooth and structured movement of the high volume of data already acquired. If it is to become a reliable platform rather than just another alternative used by a few, it will require high-performance computing and data storage expertise.

With the intensive research going on in varied fields such as IoT [47], opportunistic networks [48] and multimedia big data computing [49, 50], access to the latest technology will play a major role in the future of ICC for bioinformatics data analysis. Development of artificially intelligent cloud technologies like Amazon Web Services' Macie paves the way for further research to make cloud computing "intelligent". With the advent of next-generation sequencing (NGS) technologies, attempts to provide personalized medication through analysis of rapidly acquired genome data and the parallel development of cloud computing to deal with increasing amounts of big data, this field is of particular interest to both researchers and clients, offering great scope for further research.

REFERENCES

[1] *Domenico Talia*. Clouds meet agents: toward intelligent cloud services. *IEEE Internet Computing*, 16(2):78–81, 2012, 10.1109/mic.2012.28.
[2] W. D. Mount, *Bioinformatics: Sequence and Genome Analysis*, 2nd ed.. Cold Spring Harbor Laboratory Press, New York, 2004, 692, doi:10.1086/431054
[3] *Bioinformatics [Internet]*. 2019. Available at: https://en.wikipedia.org/wiki/Bioinformatics [Accessed: December 10, 2019].
[4] J. Xiao, Z. Zhang, J. Wu, and J. Yu. A brief review of software tools for pangenomics. *Genomics Proteomics Bioinformatics*, 13:73–76, 2015, doi:10.1016/j.gpb.2015.01.007

[5] I. Y. Abdurakhmonov. *Bioinformatics: Basics, Development, and Future*, Chapter 1, pp. 3–10. Open access peer-reviewed edited volume, doi: 10.5772/63817.

[6] L. D. Stein. The case for cloud computing in genome informatics. *Genome Biology*, 11(5):207, 2010, doi:10.1186/gb-2010-11-5-207.

[7] D. A. Benson, I. Karsch-Mizrachi, D. J. Lipman, J. Ostell, and D. J. Wheeler. GenBank. *Nucleic Acids Research*, 33:D34–D38, 2005.

[8] M. Kapushesky, I. Emam, E. Holloway, P. Kurnosov, A. Zorin, J. Malone, G. Rustici, E. Williams, H. Parkinson, and A. Brazma. Gene expression atlas at the European bioinformatics institute. *Nucleic Acids Research*, 38:D690–D698, 2010.

[9] C. Brooksbank, G. Cameron, J. Thornton. The European Bioinformatics Institute's data resources. *Nucleic Acids Research*, 38:D17–D25, 2010.

[10] T. Barrett, D. B. Troup, S. E. Wilhite, P. Ledoux, D. Rudnev, C. Evangelista, I. F. Kim, A. Soboleva, M. Tomashevsky, K. A. Marshall, K. H. Phillippy, P. M. Sherman, R. N. Muertter, and R. Edgar. NCBI GEO: archive for high-throughput functional genomic data. *Nucleic Acids Reseaarch*, 37:D885–D890, 2009.

[11] H. Sugawara, O. Ogasawara, K. Okubo, T. Gojobori, and Y. Tateno. DDBJ with new system and face. *Nucleic Acids Research*, 36:D22–D24, 2008.

[12] M. Shumway, G. Cochrane, and H. Sugawara. Archiving next-generation sequencing data. *Nucleic Acids Research*, 38:D870–D871, 2010.

[13] B. Rhead, D. Karolchik, R. M. Kuhn, A. S. Hinrichs, A. S. Zweig, P. A. Fujita, M. Diekhans, K. E. Smith, K. R. Rosenbloom, B. J. Raney, A. Pohl, M. Pheasant, L. R. Meyer, K. Learned, F. Hsu, J. Hillman-Jackson, R. A. Harte, B. Giardine, T. R. Dreszer, H. Clawson, G. P. Barber, D. Haussler, and W. J. Kent. The UCSC Genome Browser database: update 2010. *Nucleic Acids Research*, 38:D613–D619, 2010.

[14] J. Taylor, I. Schenck, D. Blankenberg, and A. Nekrutenko. Using Galaxy to perform large-scale interactive data analyses. *Current Protocol in Bioinformatics*, 10:10.5, 2007.

[15] K. Sujata, A. Rahul, and P. B. Sankar. Intelligent computing relating to cloud computing. *International Journal of Mechanical Engineering and Computer Applications (IJMCA)*, 1(1):6–8, February 2013.

[16] E. Rich and K. Knight. *Artificial Intelligence*. McGraw Hill Inc, Pennsylvania, 2006.

[17] *Hyde, Andrew Dean*. The Future of Artificial Intelligence, September 28, 2010.

[18] G. F. Lunger and W. A. Stubblefield. *Artificial Intelligence – Structures and Strategies for Complex Problem Solving*. Benjamin-Cummings, Albuquerqe, 1993, ISBN 0-8053-4780-1.

[19] Y. Hua Zhang, J. Zhang, and H. Z. Wei. Discussion of intelligent cloud computing system. In *International Conference on Web Information Systems and Mining*, Sanya, China, 319–322, 2010.

[20] J. Karlsson, O. Torreno, D. Ramet, G. Klambauer, M. Cano, and O. Trelles. Enabling large-scale bioinformatics data analysis with cloud computing. In *2012 IEEE 10th International Symposium on Parallel and Distributed Processing with Applications*, Leganes, Spain, 2012, doi:10.1109/ispa.2012.95

[21] *Amazon Elastic Map Reduce*. Available at: http://aws.amazon.com/elasticmapreduce/

[22] *InfoSphere BigInsights*. Available at: http://www-01.ibm.com/software/data/infosphere/biginsights/

[23] *Staff Report*. Microsoft Expands Hadoop on Azure, dated April 16, 2019.

[24] *European Commission*. The Future of Cloud Computing, technical report, Cloud Computing Expert Group, European Commission, January 2010.

[25] *Saratchandran Vinod*. Cloud Service Models Saas, IaaS, Paas – Choose the Right One for Your Business [Internet]. Available at: www.fingent.com/blog, [Accessed: December 24, 2019].

[26] M. Shumway, G. Cochrane, and H. Sugawara. Archiving next-generation sequencing data. *Nucleic Acids Research*, 38(supplement 1):D870–D871, 2009.

[27] H. Lee, Y. Yang, H. Chae et al. BioVLAB-MMIA: a cloud environment for microRNA and mRNA integrated analysis (MMIA) on Amazon EC2. *IEEE Transactions on Nanobioscience*, 11(3):266–272, 2012.

[28] H. Chae, I. Jung, H. Lee et al. Bio and health informatics meets cloud: BioVLab as an example. *Health Information Science and Systems*, 1(6):9, 2013.

[29] L. Ying-Chih, Y. Chin-Sheng, and L. Yen-Jen. *Enabling Large-Scale Biomedical Analysis in the Cloud*. Hindawi Publishing Corporation, London, 2013, Article ID 185679, doi: 10.1155/2013/185679

[30] J. S. Varre, B. Schmidt, S. Janot, and M. Giraud. Many core high-performance computing in bioinformatics. In *Advances in Genomic Sequence Analysis and Pattern Discovery*, L. Elnitski, H. Piontkivska, and L. R. Welch, Eds. World Scientific, 2011, Chapter 8.

[31] *Bezerra George Barreto, de Castro Leandro Nunes*. Bioinformatics Data Analysis Using an Artificial Immune Network, pp. 24–26, doi: 10.1007/978-3-540-45192-1_3.

[32] N. K. Jerne. Towards a network theory of the immune system. *Annual Immunology (Inst. Pasteur)*, 4142565: 373–389, 1974

[33] F. M. Burnet. *The Clonal Selection Theory of Acquired Immunity*. University Press, Cambridge, 1959

[34] M. Madan Babu. *An Introduction to Microarray Data Analysis*, Chapter 11.

[35] M. Armbrust, A. Fox, R. Griffith, A. D. Joseph, R. H. Katz, A. Konwinski, G. Lee, D. A. Patterson, A. Rabkin, I. Stoica, and M. Zaharia. Above the clouds: a Berkeley view of cloud computing. Technical Report No. UCB/EECS-2009-28. Electrical Engineering and Computer Sciences University of California at Berkeley, 2009, [http://www.eecs.berkeley.edu/Pubs/TechRpts/2009/EECS-2009-28.pdf]

[36] H. Zhang, G. Jiang, K. Yoshihira, H. Chen, and A. Saxena. Intelligent workload factoring for a hybrid cloud computing model. In *2009 Congress on Services – I*, Los Angeles, 2009, doi:10.1109/services-i.2009.26.

[37] A. Kouranov, L. Xie, J. de la Cruz, L. Chen, J. Westbrook, P. E. Bourne, H. M. Berman. The RCSB PDB information portal for structural genomics. *Nucleic Acids Research*, 34:D302–D305, 2006.

[38] M. S. V. Janakiram. The rise of artificial intelligence as a service in the Public Cloud [Internet]. Available at: https://www.forbes.com [Accessed: December 24, 2019].

[39] B. K. M. Sim. Towards complex negotiation for cloud economy. In *Proc. the 5th Int'l Conf. Advances in Grid and Pervasive Computing (GPC 10)*, Hualien, Taiwan, 2010, LNCS 6104, 395–406. Springer, 2010.

[40] R. Aversa et al. Cloud agency: a mobile agent-based cloud system. In *Proc. Int'l Conf. Complex, Intelligent, and Software Intensive Systems*, Seoul, Korea, 132–137. IEEE CS Press, 2010.

[41] B. Q. Cao, B. Li, and Q. M. Xia. A ServiceOriented QoS-assured and multi-agent cloud computing architecture. In *Proceedings of the 1st Int'l Conf. Cloud Computing (CloudCom 09)*, LNCS 5931, Beijing, China, 644–649. Springer, 2009.

[42] I. Lopez-Rodriquez and M. HernandezTejera. Software agents as cloud computing services. In *Proceedings of the 9th Int'l Conf. Practical Applications of Agents and Multiagent Systems (PAAMS 11)*, Salamanca, Spain, 271–276. Springer, 2011.

[43] AWS News Blog Launch – Hello Amazon Macie: Automatically Discover, Classify, and Secure Content at Scale by Tara Walker on 14 Aug 2017

[44] T. R. Coonor, N. J. Loman, S. Thompson et al. 2016 Sep; 2(9): e000086. Published online 2016 Sep 20, doi: 10.1099/mgen.0.000086

[45] K. K. Bhardwaj, S. Banyal, and D. K. Sharma. Artificial intelligence based diagnostics, therapeutics, and applications in biomedical engineering and bioinformatics. In *In the Internet of Things in Biomedical Engineering*, eds Valentina E. Balas, Le Hoang, Son Sudan, Jha Manju, Khari Raghvendra Kumar, Academic Press, Elsevier, Cambridge, MA, 161–187, 2019.

[46] Y. Yang, J. Y. Choi, K. Choi, M. Pierce, D. Gannon, and S. Kim. BioVLAB-Microarray: microarray data analysis in virtual environment. In *2008 IEEE Fourth International Conference on eScience*, Indianapolis, 2008, doi:10.1109/escience.2008.57

[47] A. Singh, U. Sinh, and D. K. Sharma. Cloud-Based IoT architecture in green buildings. In *Green Building Management and Smart Automation*. IGI Global, Hershey, Pennsylvania, 164–183, 2020.

[48] D. K. Sharma, S. K. Dhurandher, A. Kumar, A. Kumar, and A. K. Jha. Cloud computing based routing protocol for infrastructure-based opportunistic networks. In *Proceedings of IEEE India International Conference on Information Processing (IICIP), D.T.U*, Delhi, India, August 12–14, 2016.

[49] M. Devgan and D. K. Sharma. Large-scale MMBD management and retrieval. In *Multimedia Big Data Computing for IoT Applications*, S. Tanwar, S. Tyagi, and N. Kumar, Eds. Springer, Singapore, 247–267, 2019.

[50] M. Devgan and D. K. Sharma. MMBD sharing on data analytics platform. In *Multimedia Big Data Computing for IoT Applications*, S. Tanwar, S. Tyagi, and N. Kumar, Eds. Springer, Singapore, 343–366, 2019.

[51] A. Abdelaziza, M. Elhoseny, A. S. Salama, and A. M. Riad. A machine learning model for improving healthcare services on cloud computing environment. *Measurement*, 119:117–128, April 2018, https://doi.org/10.1016/j.measurement.2018.01.022

[52] A. Abdelaziza, M. Elhoseny, A. S. Salama, and A. M. Riad. Intelligent systems based on cloud computing for healthcare services: a survey. *International Journal of Computational Intelligence Studies*, 6(2/3):157–188, 2017.

CHAPTER 3

Cloud Computing Service Models: Traditional and User-Centric Approaches

Krishna Choudhary, Kshitij Gupta, Rahul Chawla, Prerna Sharma, and Moolchand Sharma

CONTENTS

3.1 INTRODUCTION

Cloud computing is defined as the use of a network of remote servers hosted on the internet to manage, store and process data, rather than using a personal computer or a local server for this task. The three prevailing service models in cloud computing are software, infrastructure and platform. Although these models offer many new potential advantages, they also offer many challenges that must be taken into consideration when choosing a solution. Areas of concern in cloud computing are data availability, security and privacy. Although service-situated architecture advances "all as a service" (acronym EaaS [1]), cloud suppliers provide their services based on various models, three of which are NIST models: Platform as a Service (PaaS), Infrastructure as a Service (IaaS) and Software as a Service (SaaS) [2]. These models are layers in a stack, yet they do not correlate. For example, SaaS can be actualized on physical machines without utilizing the fundamental dimensions of PaaS or IaaS. Conversely, a program can be run in IaaS and accessed straightforwardly, without having to incorporate it as SaaS.

Cloud computing is a way to increase capacity or add functionality dynamically without putting resources into new infrastructure, training new staff or introducing new software. Nevertheless, as more information about workers and organizations is released into the cloud, apprehensions about security are developing. Nowadays, the internet is our most fundamental way of gathering knowledge from mobile terminals. Information is uploaded to the cloud, where innovation, connection and information repositories are normal benefits given to clients over the internet. In the not so distant future, many interactions will extend beyond the present reaches of the internet and will concern the general population, the physical environment around us and virtual situations [3, 4]. As the internet advances, mobile telephones are surpassing personal computers as the most widely recognized way of accessing the web. Mobile devices have many advantages, including versatility and identification ability, and can act as pathways for a person's information. For complex data-mining operations, an integrated cloud computing arrangement and additional storage are needed. Hence the rise of a new field of research, mobile cloud computing (MCC).

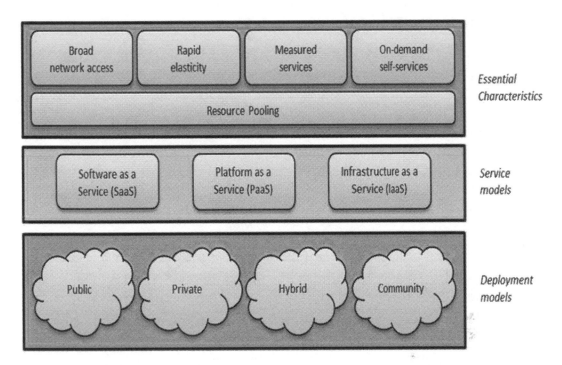

FIGURE 3.1   Definition of cloud computing according to the NIST.

Figure 3.1 shows the NIST visual model of cloud computing. While the cloud exists in the facilities of the cloud supplier, it can be operated and controlled by a commercial or government organization. It is also available for open use by the general public. This chapter will discuss traditional and client-focused service models in mobile cloud computing, consider their advantages, and examine challenges and safety scenarios.

3.2 TRADITIONAL CLOUD COMPUTING SERVICE MODELS

Picking the correct service model is an important step when deciding on cloud arrangements, and it is important to clearly understand each model's functionality and limitations. Cloud computing is a model that allows universal, advantageous and on-demand access to a shared arrangement of configurable computing assets (systems, servers, storage, applications and services) that can be conveyed and released rapidly with minimal management effort or interaction. According to an ongoing review, the percentage of companies that utilize an open cloud should increase to 51% in 2016. It is estimated that the servers sent to an open cloud will reach a CAGR compound annual growth rate) of 60%, while on-site server spending will fall by 8.6% throughout the following two years. Figure 3.2 illustrates a cloud stack.

There are three models of services in the cloud: PaaS, SaaS and IaaS. Each cloud model defines a boundary that minimizes the work a customer needs to do to organize and actualize frameworks. Each cloud service model offers a minimization and automation platform for these tasks, providing its clients with a high level of agility and enabling them to put more energy into their business problems and spend less time managing the infrastructure.

Service Models	Cloud Stack	Stack Components		Who is Responsible
	User	Login		Customer
		Registration		
		Administration		
	Application	Authentication	Authorization	Customer
		User Interface	Transactions	
		Reports	Dashboard	
SAAS / PAAS / IAAS	Application Stack	OS	Programming Language	Customer / Vendor
		App Svr	Middleware	
		Database	Monitoring	
	Infrastructure	Data Center	Disk Storage	Vendor
		Servers	Firewall	
		Network	Load Balancer	

FIGURE 3.2   Cloud stack.

3.2.1 Types of Traditional Cloud Computing Service Models

3.2.1.1 Infrastructure as a Service

IaaS is defined as the use of server, repository and virtual cloud to provide utilities. The internal architecture covers installation, communication systems, computer centres and the organization of the virtualized handling of services controlled by a service supplier. The objective of the service includes components within the client's area of control, as well as virtual machines and their working frameworks [5]. Figure 3.3 shows an IaaS block.

According to Oppenheimer, automatic facilitation of the framework could be advantageous for bandwidth and to calculate heavy loads. The opposite seems to be true for less and discontinuous usage [6]. Other reasons to take into consideration when deciding on a facilitated arrangement include security, management and reallocation of data storage.

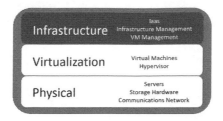

FIGURE 3.3   IaaS block.

Most organizations have work and information from different companies, increasing security and privacy risks. For instance, when virtualization is used, the hypervisor is given access to the hardware. But a framework administrator may not be given access to the visitor framework operating inside a client's virtual machine. Some hypervisors have a special domain that can be used to access the virtual server machine's active memory [7]. If a hypervisor is susceptible, it may be able to catch the substance of the memory, the traffic of the virtual system, and different forms of communication happening inside the control domain [7].

The careful assignment of jobs to staff, such as logging and applying security guidelines with the least benefits, is a good start when dealing with challenges to platform security. It could also be valuable to eliminate or alter reasons for special domains in order to mitigate the spying hazard from the hypervisor. From a communication organization point of view, scrambling data secured inside the visitor operating framework by, for example, using a virtual private network, may help to insure against any present or upcoming vulnerabilities in the virtual layer [7].

An IaaS case described by Toews et al. is relevant in this context. Their investigation included an attempt to manufacture a half-and-half cloud network for an international physicists' gathering. The researchers expected access to the whole system for configuring, creating and emulating various models. The team built up an IaaS service model-based cloud infrastructure. For their framework, software compatible with Piece Virtual Machine (KVM) and the Amazon Web Services (AWS) application programming interface [8] called Eucalyptus was chosen. Cloud computing hubs were geographically dispersed around the world to reduce inactivity and enhance the end user's experience. There was no requirement for individual work stations, since the hardware was assembled into sharable gatherings. Rather, occasional acquisitions were to the whole cloud asset arrangement. The physicists did not have to maintain server hardware because cloud assets were now managed centrally by many organizations. Therefore, the researchers could access the various assets to carry out their simulations rather than handling actual systems. On the off-chance that these assets turned out to be less than required, they had an option to buy Amazon's web services (AWS). Several administrative tools were also created, including "a GUI based web application with an interface supporting assistant" [8], which made the IaaS service resemble previous versions and facilitated the utilization of the cloud. Only two explicit operating frameworks were run in the virtual machines to streamline administration and provide consistency. An issue with the hardware was the availability of IPv4 addresses, which constrained the number of virtual machines accessible on the internet. With increasing demand and popularity for services, it was noticed that they eventually experienced a problem of inordinate memberships due to limited IPv4 addresses. They eventually decided that Eucalyptus "was not appropriate for" their cloud situation and reevaluated several existing frameworks to find something more suitable [8].

Cloud computing enables association between dissimilar individuals and work gatherings. It has conquered the challenges arising from existing arrangements. It has enabled productive admittance to massive IT assets and rearranged administration through web applications and remote access.

3.2.1.2 Platform as a Service

PaaS provision demonstrates that programming languages, APIs and advanced middleware can be accessed that permit endorsers to create conventional applications without the need to install or arrange improved conditions. PaaS is based on IaaS and offers quite a lot of similar advantages, such as hardware virtualization, dynamic asset distribution, utility computing and reduced investment costs. Utilizing the tools incorporated in the cloud, engineers can create applications and facilities that benefit from virtualized hardware, data dismissal and high accessibility. When improvements are completed, the application can be conveyed to clients through the internet [9]. Examples of PaaS suppliers are Google App Engine, Microsoft Azure and SalesForce.com [10].

Compatibility is an important task performed by PaaS. Features such as languages, software or intermediate software are common to all PaaS suppliers [11], making choice difficult [12]. Cloud-controlled podiums are classified as either total or partial PaaS. Complete PaaS enables the customer to create a complete arrangement via an online UI without the engineer having to install anything except an internet browser. Partial suppliers provide the client with a lot of tools as a service, while at the same time requiring them to install applications and create arrangements on their devices. Full PaaS is chiefly susceptible to compatibility matters and possible provider blockage, but requires less in the way of management and upkeep, and is ready for immediate use.

An additional concern for PaaS clients is that software engineers continue to be cautious about new stages and learning novel APIs. These concerns should be alleviated after some time as suppliers acquire a reputation and popularity. At the moment supplementary APIs (OpenShift [13], Cloud Foundry [14]) can be used to create independent apps for platforms, allowing the client to pick the cloud supplier to be used for its application.

As in IaaS, safety is an important issue when working with a PaaS model. Open clouds limit clients' ability to secure their registered data in the way they might with a corporate organization, and to control where it is stored [15]. The PaaS supplier must ensure there is no way for a customer to access the platform, arrange the traffic or access data of another client.

Figure 3.4 shows the magnetotelluric (MT) inversion calculation process, a technique utilized by geophysicists to characterize underground structures [16]. Utilizing the assembled data and a level of variation, researchers can create 3D mapping of the underground structures with software and algorithms, providing a more accurate estimate of the position of geothermal supplies than was possible with previous techniques. However, MT algorithms can take a long time to run on a desktop PC with a single processor, especially when a data set is frequently executed multiple times, adjusting the variables to obtain increasingly advanced mapping. The abundance of data, together with the multifaceted nature of speculative MT calculations, needs a large number of computing assets. According to Mudge, "The MT's requirement of RAM's quantity can be recalculated by squaring the current size of the model. Signification is that the generation of a continental model with a huge number of cells will need RAM on board" [17]. Researchers have noted the viability of using IaaS with parallel computation to reduce the time taken to compute MT mapping, as well as the advantage of undertaking activities from distant sites over the internet. By

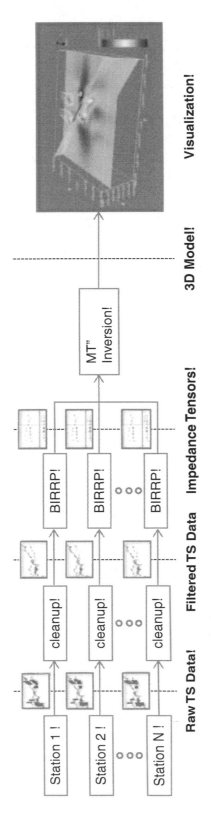

FIGURE 3.4   MT inversion calculation process.

having the capacity to track the program in parallel and apply the IT assets of IaaS, the investigators achieved a reduction in overall handling time by a factor of four, meaning that the outcomes could be obtained in days instead of weeks [16].

This case confirms that cloud services models represent an excellent substitute for a patented computing organization that involves irregular computation and high asset needs. Though problems can arise with compatibility and learning novel APIs, these concerns are similar to those that accompany application upgrades. Although the engineers of the MT venture were new to cloud architecture and were not familiar with current procedures, they could achieve their goals in a convenient way, obtaining excellent results.

Security concerns for engineers on cloud platforms are similar to those pertaining in traditional environments where outsiders are admitted; designers are in charge of creating secure applications, and applications must be tried, preferably by outsiders. The whole PaaS model is responsible for a significant part of the security issues associated with building up an answer for the cloud supplier. Most cloud-explicit security and privacy apprehensions must be the obligation of the provider.

3.2.1.3 Software as a Service

Software as a Service (SaaS) provides supporters or paid clients with software and services that reside in the cloud and cannot reside on the client's device. A user of a SaaS app needs minimal client software, for example, an internet browser to access the facilitated application in the cloud. This reduces hardware requirements for end clients and enables centralized software control, arrangement and maintenance. Gmail [18], Hotmail and Google Apps [19] are examples of popular SaaS applications.

SaaS has numerous advantages for organizations in terms of financial planning and cost savings. According to Microsoft, the greatest advantages of executing applications utilizing a SaaS model are that minimal investment in hardware, software and staff is needed up front [20]. An investigation by Hurwitz and Associates found that SaaS arrangements saved 64 per cent over four years compared with a local arrangement [21]. An investigation led by Hurwitz and Associates concluded that SaaS arrangements presented a 64 per cent saving on a comparable local arrangement over a four-year period [21].

SaaS arrangements raise numerous concerns over the security of company data. System World alludes to regulations such as the federal law on the management of information security, which expects customers to maintain confidentiality of data within the nation [22]. While access to data anywhere is advantageous and decreases the requirement for company representatives to keep confidential information with them, an unsafe endpoint may turn out to be a major hazard. Any company considering SaaS must carry out detailed scrutiny of a proposed provider's arrangements. Messmer asks, for example, about the SaaS representatives that have access to the root database. Are the data kept scrambled? Are customer data separated? Which information is acquired in the control archives [23]? Knowing the provider's security strategies will be crucial in the leadership process.

An example case is that of San Francisco University (USF) and its acceptance of an SaaS arrangement. USF is the oldest university in the city and comprises six campuses with 1300 members of staff and currently more than 8500 undergraduates. The university supports more than 1000 laptops, both on its premises and mobile [24]. USF established a series of requirements related to data security. Their first proposal to the end client was to copy the security of mobile devices. Among USF's users are large data creators, many of whom spend much of their time on their PC producing intellectual property that is required to be secured. California's law requires all significant parties to be advised in the event of a data breach. It is inconceivable that if a laptop is stolen, the data it contains could not be recovered. Disaster recovery procedures were also necessary. The history of earth tremors in the region underlined the requirement for dependable off-site security backup. A financially advantageous arrangement was another necessity [24]. The current multisystem arrangement had many drawbacks, including mind-boggling expense, high administration overheads and a lack of satisfactory data security for mobile end clients.

USF's IT department sought internal answers to these requirements, but despite these efforts, only a small minority of their clients were able to benefit. Furthermore, this internal arrangement was not able to meet the never-ending rates of data development [24]. With ongoing advances in SaaS, USF staff started to look at cloud-related arrangements. Mozy was their chosen supplier for SaaS data security facilities. Mozy provides cloud insurance services for servers, laptops, PCs and mobile devices. Mozy utilizes a pay-per-use model that enabled USF to appreciate the advantages of a SaaS arrangement that only pays for what it needs, with no investment expenses. Mozy has five databases that allow for adaptability amid data repetition and data position [24].

A company that moves to a cloud-based arrangement can expect numerous tests and advantages. In this case, by actualizing a data assurance arrangement based on SaaS, the organization enjoyed numerous normal rewards, together with price savings, rapid implementation, low capital costs and expanded capabilities. With SaaS, it is also in a position to absorb any amount of data development. This case was a genuine instance of an effective real-world implementation of cloud-based computing, having established its capabilities.

3.2.2 Security Issues in Traditional Cloud Computing Service Models

Nowadays, small and medium enterprises (SMEs) are beginning to understand that simply by accessing the cloud, they can achieve rapid access to top commercial applications or dramatically increase their organization's assets, without regard to expense. Gartner characterizes cloud computing as "a PC style where enormously scalable IT-enabled abilities are conveyed 'as a service' to outside customers utilizing Internet innovations". Cloud suppliers are now seeing a wide opening in the marketplace. The cloud has many advantages, for example, rapid and universal implementation and provisioning, low cost, increased resistance, greater assurance against system threats, minimal-effort disaster recovery and data repository arrangements, control of security controls and real-time manipulation of the framework. Cloud computing helps applications and databases to focus on large amounts of data without having to consider data management and services. This extraordinary characteristic presents numerous novel security challenges, including but not limited to accessing susceptibilities, virtualization susceptibilities or web app susceptibilities, for example, SQL infusion and XSS (cross-site scripting), confidentiality and problems from outsiders.

Even though the cloud aims to provide improved use of assets through virtualization techniques and to take over much of the burden from the user, the security hazards of a cloud platform remain unpredictable (Figure 3.5).

In Figure 3.5, the first layer shows the individual cloud arrangement models: private, network, open and hybrid cloud organizations. The layer directly above the organization layer explains the different conveyance models used inside an arrangement model: SaaS, PaaS and IaaS. The cloud security arrangements vary according to the arrangement model that is used, the method by which it is carried out and the atmosphere it creates.

3.2.2.1 Security Problems of SaaS

The client is dependent on the supplier for suitable security measures in SaaS. The supplier's main job is to prevent clients from viewing each other's data. Thus, it is difficult for the client to be certain that the right security measures are in place, and that the app will be available whenever needed. With SaaS, cloud users will replace old software applications with new ones from an endless central supply of applications, while protecting or enhancing the security functionality of the legacy application and satisfactorily migrating the data.

The SaaS software merchant may host the application on a private server or send it to a cloud computing infrastructure service provided by an outsider supplier (for example, Amazon or Google). The use of the cloud combined with the pay-as-you-leave (develop) approach reduces interest in infrastructure services and allows the application to focus on giving improved services to clients.

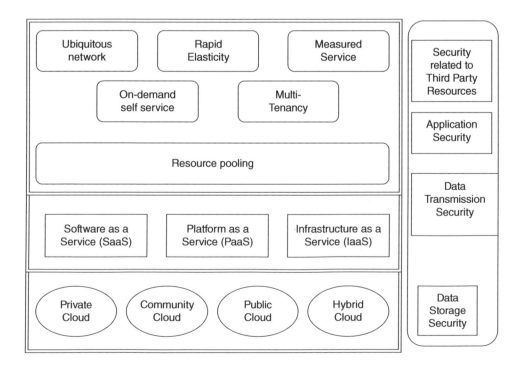

FIGURE 3.5 Complexity of cloud platform security.

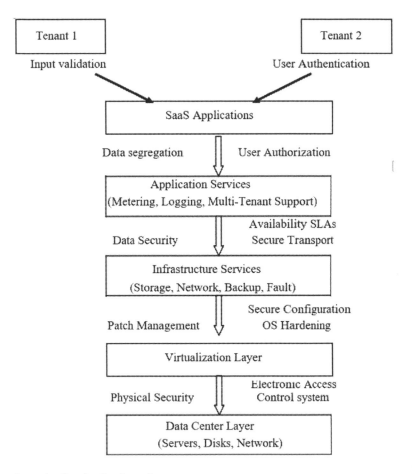

FIGURE 3.6   Security for the SaaS stack.

Organizations today protect corporate data (records, transactions, estimates, and so on) themselves. This is done by access control. In the SaaS model, an organization's data is kept with the SaaS supplier's data, together with that of various other organizations. If the SaaS supplier is exploiting an open cloud service, the data may be kept together with other unrelated SaaS applications. The cloud supplier may also duplicate data at various places across different countries to enable a high level of data accessibility. Many problems are caused by the absence of information on how the data are kept. There are some genuine concerns about application susceptibilities, data breaches and accessibility that have financial and legal implications.

The following key security components should be carefully considered as an integral part of the SaaS application and organizational process (Figure 3.6):

- Data security
- Network security / data locality
- Data integrity

- Data segregation / data access

- Authentication and authorization

- Data confidentiality

- Web application security data breaches

- Virtualization, vulnerability and availability

- Backup

- Identity management and sign-on process.

3.2.2.2 Security Problems of PaaS

The cloud supplier gives the clients control for their engineers to build applications using the platform in PaaS. PaaS offers more capabilities than SaaS because of features that are customer ready. This tradeoff also applies to security features, where the inherent capabilities are incomplete, yet there is greater flexibility to add further security layers. All security matters below application level, for instance, counteracting host and system disruption, should be the supplier's responsibility, and the supplier must offer rock-solid guarantees that data remain inaccessible in applications.

In order to leverage an enterprise service bus (ESB), applications need to be designed in such a way as to anchor ESB straightforwardly using a protocol like Web Services (WS) Security (Oracle, 2009). The efficiency of the security applications has to be calculated exactly, for example, vulnerability scores and patch coverage can be used to indicate the quality of application coding. Attention should also be paid to how malicious codes affect the working of new cloud applications. Attackers not only run malicious code inside the client's environment but can also perform widespread black-box testing and penetrate the infrastructure.

3.2.2.3 Security Problems of IaaS

Although many virtual machines can handle some security issues, there are still many that cannot be handled using a virtual machine and there is no security feature for the virtualization manager. Another factor to be considered is the reliability of the data that are present inside the supplier's hardware. With everything related to information security treated as virtual in cloud technology, ultimate control over data becomes almost impossible for the data owner. A combination of several systems is needed to achieve the maximum level of security in cloud systems.

Cloud service models decide upon the security duties of both the supplier and the purchaser. For example, Amazon's Elastic Compute Cloud (EC2), an IaaS model, assigns control over physical, environmental and virtualization security to the merchant, and control over security of the IT framework, which includes the OS, applications and data, to the shopper.

Cloud is designed to work using the internet; therefore, all the security-related issues present on the web are also a part of cloud infrastructure. IaaS is susceptible to numerous security problems, depending on the cloud organization model through which it is being

TABLE 3.1 Cloud Service Deployment Models

	Infrastructure Management	Infrastructure Ownership	Infrastructure Location	Access and Consumption
Public cloud	Third-party provider	Third-party provider	Off-site	Untrusted
Private/community cloud	Organization or third-party provider	Organization or third-party provider	On-site or off-site	Trusted
Hybrid cloud	Both organization and third-party provider	Both organization and third-party provider	Both on-site and off-site	Trusted and untrusted

shown. An open cloud is subject to major hazards, whereas a private cloud appears to be safer. The physical security of the infrastructure and any disaster management are also very important because infrastructure affects not only the hardware where data is kept, but also its communication path.

Though cloud infrastructure is rapidly developing, we may expect some new advancements in the field. Cloud enables purchasers and suppliers to be present at various places and virtually control assets on the internet. In the cloud, data is generally communicated through countless outsider organization devices from source to destination. Although cloud frameworks use the same internet protocols that are used on the internet, the cloud still has high-end requirements. Inside the cloud, strategies and protocols must be arranged to ensure proper transmission of data. The interruption of data caused by non-cloud users through the web is also a major concern. The intricacy associated with IaaS because of the service organization models is demonstrated in Table 3.1.

3.2.2.4 Current Security Solutions

Much research has been done on cloud security, and numerous organizations are keen on creating appropriate security arrangements. The Cloud Security Alliance (CSA) procures arrangement suppliers, non-benefits and consultants. The Cloud Standards site collects and coordinates data about cloud-based standards. The Open Web Application Security Project (OWASP) deals with the top susceptibilities of cloud-based or SaaS models and is updated as the threat landscape changes. The Open Grid Forum allocates reports for regulating security and infrastructural specifications and information for lattice computing engineers.

The best way to provide security to web applications is by building and improving the existing security architecture. Tsai et al. (2009) define a four-level framework for electronic improvement. Berre et al. (2009) provide a blueprint for cloud-driven improvement, and the X10 language helps acquire better utilization of cloud capabilities such as large-scale parallel handling and simultaneity (Saraswat Vijay, 2010).

Krugel et al. (2002) proposed increasing the yield of a packet "sniffer" to explicit services as a way of addressing security issues derived from irregular packets coordinated to explicit ports. One solution that is generally ignored is to close the vacant services, have patches updated regularly and reduce the authorization privileges of clients and applications. Raj (2009) proposes asset segregation to guarantee data security, with the processor's cache in virtual machines segregated from the hypervisor cache. Hayes (2008) argues that there is no way of knowing whether a cloud supplier has legitimately erased a user's cleansed data, or kept it for ambiguous reasons.

According to Hayes (2008), "Making an outsider service to handle personal archives increases awkward inquiries concerning control and possession: If one moves to a contending provision supplier, can you carry data with you? Would one be able to misplace access to reports if one fails to wage a bill?" Subjects of discretion and control cannot be established, yet just guaranteed with limited service-level agreements (SLAs) or by ownership of the private cloud itself. Another arrangement proposed by Milne (2010) as a widely used solution for US organizations is "private clouds".

3.2.3 Summary

As we have seen in this section, cloud computing is widely accepted by big organizations and offers many advantages with centrally managed provisioning and remotely available applications. It is important to decide whether cloud technology is a suitable business arrangement, and if so, which model provides the best balance of control against limited hardware, outlining and upkeep costs. Some organizations have profited from the reduction in principal costs and have focused on ensuring their data is secure at the expense of data severance and duplication concerns. The ongoing challenges associated with these models will, however, eventually be fully addressed by continuing innovation. The integrity, security and privacy of cloud computing is of prime importance.

However, there remain many real-world issues with a cloud framework. Many of the problems concern SLAs, security, privacy and power usage. Security remains a deterrent for many customers who, until a well-structured security module is implemented, are unable to understand the advantages of the cloud. Each component of cloud ought to be examined at the higher and smaller-scale level. Without a cohesive arrangement for every cloud component, both large-scale and small-scale, to convince the customer, the prospects for the cloud remain dismal.

The present research aims to establish a cohesive and progressive security model taking account of various dimensions of data security for specific cloud infrastructures. The research addresses application and data privacy in the cloud, with a security philosophy that varies according to different transactions. Security in everyday life relies upon necessity and costing. A cloud that has a typical security procedure defined is vulnerable to hackers since hacking the security framework would put the whole cloud at high risk. Providing modified security as a service to applications therefore makes sense. This research targets the many practical concerns regarding dynamic security and data storage based on meta-data information, providing a practical framework.

3.3 A USER-CENTRIC APPROACH TO A MOBILE CLOUD COMPUTING SERVICE MODEL

Mobile phones are replacing personal computers for internet access due to their cutting-edge features, including portability, telecommunication and detection capabilities. Mobile devices are rapidly becoming major service participants. Existing traditional mobile cloud computing service models are being replaced with new client-driven mobile cloud computing services. Traditional client–server-based mobile service models are unable to meet the growing demand from mobile clients for variety, client knowledge, security and privacy. Cloud computing allows mobile devices to dispense with the complex operations of mobile apps that are unusable on mobile devices alone.

In mobile cloud computing (MCC), a mobile element is imagined as either a hardware mobile device or as mobile software inside a cloud-enabled framework. In the virtualized cloud framework, the key feature of a software agent is its versatility linked with its software codes. Interruption, vitality utilization, proximity, information-hiding capabilities, systems administration, communication availability, data security and sharing are all properties of MCC that are able to generate a world of physically arranged frameworks and virtualized substances plotted to the physical frameworks, safeguarding and, at times, expanding their capacities and abilities. MCC research focuses on tight communication between development and the combination of the digital-physical framework (CPS) and virtual digital framework (CVS), in which the CPS is largely created by physical smart and moveable substances, and the CVS is primarily shaped by cloud-based assets and services.

3.3.1 Current Mobile Cloud Services Models

Current internet cloud services are largely divided into three service models, according to their virtualization layers: Infrastructure as a Service (IaaS), Platform as a Service (PaaS) and Software as a Service (SaaS). Due to the contribution of CVS and CPS, MCC service models are especially distinguished by their computational fundamentals, where the arrangement of MCC service models may operate jobs between mobile elements and their fabricated cloud-provisioned assets. Based on this interpretation, current MCC services are of three types: Mobile as a Service Provider (MaaSP), Mobile as a Service Consumer (MaaSC), and Mobile as a Service Broker (MaaSB). These models are illustrated in Figure 3.7, where the arrows depict service-handling streams from suppliers to beneficiaries.

MaaSC is based on the old-fashioned server–client model, presenting virtualization and other cloud-infrastructure advances in its initial stages. Mobile devices in the cloud can re-appropriate their computational and storage capacities to achieve better performances. In this architecture, communication is only one-way, i.e., from the cloud to mobile devices. Generally, all currently working MCC services fall under MaaSC. MaaSP is slightly different. Under MaaSP, a mobile device uses sensors (GPS, camera, spinner, and so on) to detect data from its environment and then pass information to further mobile devices using the cloud. The services specified for mobile devices are variously built depending upon their detecting and handling abilities.

The extension of MaaSP can be assumed to be MaaSB. MaaSB gives systems administration and data-rendering services to different mobile devices. MariaDB, one of the most popular open-source relational databases, is wanted for a few situations because mobile devices typically have constrained detecting ability linked to sensors that are designed for specifically planned functionalities and for detecting places. For example, they can be utilized to gather clients' physical actions using wellness bands. MaaSB extends the cloud boundaries to mobile devices as well as remote sensors.

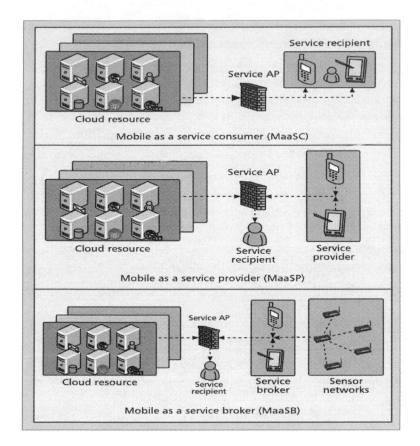

FIGURE 3.7   Mobile cloud computing service models.

3.3.2 Existing Mobile Cloud Applications

Currently available MCC services and their respective applications are summarized in Table 3.2. Every service is characterized in one or more different service models. The most widely used MCC service model is MaaSC, as generally existing mobile devices are restricted due to computation and vitality limitations.

Table 3.2 shows the four MCC types along with their respective approaches, further categorized into service models.

3.3.2.1 Mobile Cloud's Computational Task

Asset-escalated computation is challenging for mobile devices, depending on the cloud arrangements. Dividing computation tasks between mobile devices and clouds is extremely wasteful, taking a number of performance dimensions into account, for example, vitality utilization, CPU usage and system delay. It is the main task of MCC to choose the most efficient method of loading computational tasks on the cloud. Clone Cloud [26] and MAUI [27] are the main examples in this field.

TABLE 3.2   MCC Applications and Services in Detail

| MCC Service Types | MCC Services and Applications | Service Models | | |
	Representative Approaches	MaaSC	MaaSP	MaaSB
Mobile cloud computation	CloneCloud [7]	✓		
	MAUI [8]	✓		
	ThinkAir [9]	✓		
Mobile cloud storage	Dropbox, Box, iCloud, GoogleDrive and Sky Drive [10]	✓		
	WhereStore [11]	✓		
	STACEE [12]	✓	✓	
Security and privacy	CloudAV [14]	✓		
	Secure web referral services for mobile cloud computing [13]	✓		
	Zscaler [15]	✓		
	Google Wallet [25]	✓		
Context awareness	An integrated cloud-based framework for mobile phone sensing [16]		✓	✓

3.3.2.2 Mobile Cloud Storage

Storage is a major concern for mobile devices. Many storage services exist, for example, Google Drive, iCloud, Box, Dropbox and Sky Drive [28]. Where Store [29] is a prominent location-based data storage service that uses separated replication of the device's location history. STACEE [30] proposed distributed point-to-point cloud storage where mobiles, set-top boxes, tablets and organized repository devices all act as storage devices.

3.3.2.3 Security and Privacy

CloudAV [31] promotes a cloud-based security model for threat identification. Web referral secure services [2] enable anti-virus and anti-phishing facilities using the cloud. To prevent mobile clients from opening phishing sites, recommendation services rely upon a protected search engine to authenticate URLs that are opened using mobiles. Zscaler stands out among profitable cloud-based security firms with its secure internet access intended for mobile devices.

3.3.2.4 MCC Context Awareness

Nowadays, smart mobile devices act as an information gateway for mobile clients, who use them for various personalized activities, such as checking emails, making appointments, using the web, looking up places of interest, and examining personal behaviour data built on data mining and ML. For example, in [30], every mobile device has its own dedicated mobile cloud motor (MCE) with three modules: distributed buy-in module, choice module and setting awareness module.

3.3.3 From Internet Clouds to User-Centric Mobile Clouds

3.3.3.1 Current Mobile Cloud Computing and the Direction of Travel

Current MCC service suppliers are characterized by the service provided. Most current PC models are similar to old-style client–server service models. The following issues relate to current MCC services.

- **MCC symmetric service model:** Most MCC models are not symmetric. As Table 3.2 shows, mobile devices are generally viewed as customers of cloud computing services. Service management (for example, PCs and storage services) is generally unidirectional: to mobile gadgets from the cloud. As the capability of mobile devices increases, they can become more collaborative in applications.

- **Personalized learning:** Today's mobile cloud computing generation has ample data sources. A single data source is not enough to support MCC applications. Furthermore, data gathered through different systems may be without any structure or classification. For instance, various system interfaces and services are present for the client's device (a remote sensor that organizes social systems, vehicle systems, personal and body area systems, and so on).

- **Trusted model focused on the client:** Most existing cloud trust models are centralized, with every mobile element trusting the cloud service supplier. Storage of personal data is a major obstacle for most of the mobile applications on the cloud. A fiduciary management structure should be created inside the virtualized cloud framework to deal with clients' private business. In the virtual world, assets are facilitated that adapt to the client's requirements.

3.3.3.2 The User-Centric Approach to Mobile Cloud Computing

A strong combination of physical and virtual capabilities is required for cutting-edge MCC applications performed on mobile devices and cloud respectively. Also, because of the versatility of mobile clients, MCC application capabilities do not support the hosts on which they run [32, 33]. Figure 3.8b provides an illustrative example of vehicle traffic management, where video capture (VC) of different vehicles (dashed line) or combined centralized video is required for a comprehensive view of an entire road junction. Here, VC suppliers are mistaken and chosen based on their coordinates. Besides, VC work cannot be utilized just for a solitary vehicle; it may also be utilized for the management of traffic on the road, the discovery of accidents/hazards, and so on.

3.3.4 Design Principles of User-Centric Mobile Cloud Computing

MCC operates like any other service model, with mobile devices and their linked assets acting as mobile clouds enabling computing, storage, data collection and dissemination, and systems administration. For client-based MCC, adjustments are made to traditional internet clouds in accordance with the following principles:

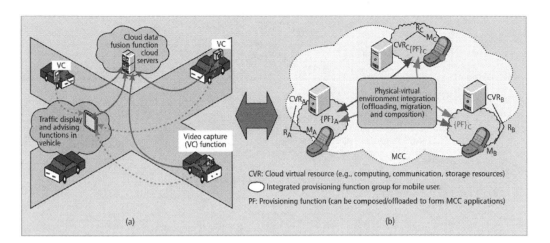

FIGURE 3.8 Mobile cloud applications: (a) MCC application scenario. (b) A user-centric MCC application model.

- **Principle 1:** Focus on the client. MCC applications must be structured in such a way that control of data and activities go hand in hand with privacy and security for the client. Cloud assets must be gathered and assigned to individual clients based on custom mobile applications.

- **Principle 2:** An application platform that is service situated. Because of the symmetrical service model, each mobile hub may theoretically act like an MCC service supplier; hence, MCC naturally uses a service-arranged application platform.

- **Principle 3:** Versatility. Client centre assets are dynamically assigned and controlled based on mobile cloud application requirements.

- **Principle 4:** Portability. MCC applications should be limited to maximize proficiency, by utilizing a lot of framework performance measurements.

- **Principle 5:** Virtual representation. MCC maintains reliable virtual representation that is available to each client. Virtual representation helps mobile clients to perform actions, such as tracking day-to-day activity to create client behaviour and activity profiles.

3.3.5 Mobile as a Representative: A User-Centric Methodology

In addition to the service models introduced above (for example, MaaSP, MaaSC, and MaaSB), there is another client-focused cutting-edge MCC service model termed mobile as a representative (MaaR). MaaR's architecture is shown in Figure 3.9. MaaR allows every client to be spoken to using a virtualized substance in the cloud with the help of its physical element (a mobile device). Client actions and characteristics are gathered from the physical

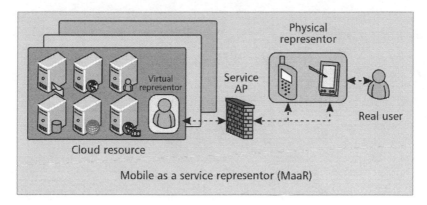

FIGURE 3.9   Mobile as a representative (MaaR).

world and can be compared with virtual elements in the cloud for examination and handling. In MaaR, clouds and mobile devices are profoundly collaborative. MaaR is able to perform functions that are difficult in the present MCC architecture, thus making mobile substances perform more effectively.

3.3.6 User-Centric MaaR Model-Based Application Scenario

The projected future MaaR model is illustrated in the example of video collaboration and vehicle location introduced in Figure 3.8a. Alice's mobile phone uses on-board sensors to determine its position and speed and record images from the road. Using this information, a virtual representation of the mobile device can be created in the cloud for Alice. The essence of MaaR is that the virtual representative speaks to the actual situation of the physical client. In practice, representatives are actualized using a progression of software (for example, OSGi packages) in a dedicated virtual machine assigned to Alice, who is able to choose the data that is communicated and secured. The dedicated virtual machine is the proprietor of the application. In the model, the virtual machine may be facilitated in an open or dedicated cloud, as Alice chooses.

In the example, client Bob wishes to determine the present state of traffic in a radius of 5 miles from Alice. Clients can receive MaaR on their mobile devices which may offer tracking capabilities (for example, GPS, video/camera), which Bob can search for, so Bob's visualization capacity can call these capacities in real time via point-to-point associations. MaaR also depicts social diagrams for every client. For example, if Bob is driving in the area in the middle of the day, Bob's MaaR service representative can recommend nearby restaurants highly rated by Bob's associates. Different recommendations relevant to Bob's day-to-day activities and work capacities based on his present position can be rapidly called up when Bob requires them. The personal proposals are designed based on the correlation between the position and various data gathered by the MaaR service representative. Figure 3.10 shows the MaaR conceptual architecture.

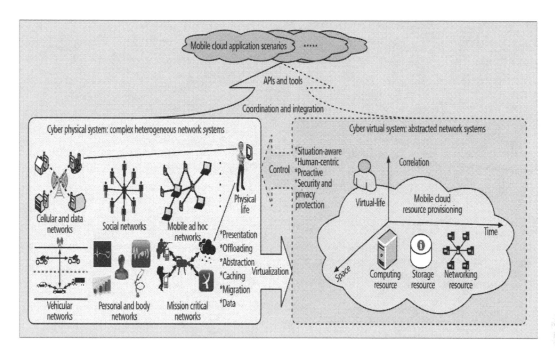

FIGURE 3.10   MaaR conceptual architecture.

3.3.7 Summary

This section has provided an overview of current advances in client-focused mobile cloud computing, together with a classification and representative examples of current MCC service models. The transformation of the traditional internet cloud into the client-focused mobile cloud is also analyzed, with the issues this poses for current customer focus. Client-focused MCC is introduced with a descriptive example of an MaaR service model.

3.4 CASE STUDIES

3.4.1 Instagram: The Billion-Dollar App

In 2010, the picture-sharing app Instagram got 25,000 sign-ups on its inaugural day. Three months after launch, it boasted a million clients, rapidly rising to ten million and, 18 months later, to 30 million. Instagram's initial version was an iOS, so it only had iPhone clients. When the Android version was finally launched, they gained a million clients in a single day. In April 2012, less than two years after launch, Facebook purchased Instagram at an estimated value of US$1 billion.

This story features the ability to calculate assets upon demand. These talented designers have managed to fabricate an astonishing and exceptionally scalable architecture in a brief timeframe. If they had attempted such a rapid climb with a focal point of physical data, they would never have been able to purchase the hardware fast enough to stay abreast of development. Thanks to the cloud, they didn't have to administer data focuses or acquire, install and control hardware. Instead, they concentrated on application building and client encounters, two things they were known for. For start-up organizations with a focal point of existing data, cloud technology is a gift.

3.4.2 Netflix: An Established Company That Shifted Its Entire Enterprise to the Cloud

Netflix is a giant in the transmission of video content on the internet. In 2009, one per cent of all consumer traffic was achieved using our focal point of Netflix data. By the end of 2010, a significant proportion of this traffic had shifted to AWS, Amazon's open cloud arrangement. Netflix's 2013 target was to have a 95 per cent share of all services, including operational and customer traffic services. In its innovation blog, Netflix stated that its main reason for using the cloud platform was the gigantic quantity of inbound traffic. They concluded that they would prefer to concentrate on developing and enhancing their business applications (the key capabilities of Netflix) and allow Amazon to concentrate on the infrastructure (the main capability of AWS). Companies that create local arrangements must purchase over-capacity to manage the peaks. Netflix's solution to the problem of anticipating demand was to control assets on demand from the general population cloud and concentrate on creating automatic scaling.

Cloud environments are ideal for horizontally scaling architectures. They don't have to guess about the future needs of network, storage or hardware. They can programmatically access these shared pools instantly. Netflix also sees leveraging cloud computing as a competitive advantage. The company is able to scale at amazing levels while controlling costs and reducing the risks of downtime. (Netflix innovation blog, 14 December 2010)

3.4.3 NOAA, Email, and Collaboration in the Cloud: A Government Case

The National Oceanic and Atmospheric Administration (NOAA) is a centralized US agency with more than 25,000 employees whose main aim is to predict and study changes in the climate. Employees work on land, in the air and at sea, and rely mainly on internet devices and on teamwork with individuals and different agencies. In early 2012, NOAA changed to a cloud-based email system, Google's Gmail, which permits proficient email and association abilities, prompt messaging and video conferencing using the cloud. Migration to cloud technology helped NOAA reduce its costs by half. NOAA's administration claims that cloud-based email and its tools are quick and are easier to actualize than local arrangements, providing a superior overall service at half the cost and with less effort.

3.4.4 The Obama Campaign: A Not-for-Profit Case

The sort of challenges faced by the Obama campaign technical team are extremely rare. They had to rapidly create a series of applications including a web-based business fundraising platform that could handle more than a billion dollars and last just half a year, experienced a colossal increase in the final stages, and then made a backup of everything. The team depended mostly on cloud technology arrangements and utilized the facilities of each service model (IaaS, PaaS and SaaS). The telephone application reached 7,000 simultaneous clients at its peak on election day. They spent about US$1.5 million on web facilitating and web services, of which, astonishingly, more than US$1 million went to a local company. The facilitating company managed social systems and advertising, while the remaining over 200 applications were kept running on under US$ 500,000 using cloud technology.

3.4.5 Summary

Cloud is one of the greatest technological changes since the introduction of personal computers and the wide dissemination of the internet, and has made great advances in recent decades. But cloud computing has only just started building. Initially, only start-up companies and small organizations incorporated cloud, but since early 2013, cloud computing has become widely accepted, and corporate spending plans for the cloud are developing at a colossal pace. Like anything that is so recent and underdeveloped, the cloud also lacks standards and proper implementation. Despite many interruptions, the overall performance of cloud service suppliers has improved through the years as their items and services have matured, resulting in extraordinary examples of companies like Netflix and Instagram overcoming adversity to end up with consistent progress. Commercial software licenses and hardware are losing money. The key decision for companies is choosing the appropriate cloud solution to tackle specific business issues. An understanding of the cloud services models (SaaS, PaaS and IaaS) is essential for companies taking an interest in the cloud.

3.5 CONCLUSION

With the advances in cloud technology, clients experience the seamless service arrangement of cloud computing. Clouds have never been a client-driven innovation, and clients have to be totally subject to service suppliers. With today's exponential development, the novel idea of clients self-organizing to create their own clouds has arisen. This architecture, based on precept and personality management ideas, is referred to as the client-driven cloud model.

REFERENCES

[1] Microsoft. *Software as a Server (SaaS): an enterprise perspective.* [Online]. Available at: http://msdn.microsoft.com/en-us/library/aa905332.aspx [Accessed February 16, 2012].

[2] P. Stued, I. Mohomed, and D. Terry. Wherestore: location-based data storage for mobile devices interacting with the cloud. In *Proceedings of the 1st ACM Workshop Mobile Cloud Computing & Services: Social Networks and Beyond*, San Francisco, CA, 2010.

[3] S. Rajan and A. Jairath. Cloud computing: the fifth generation of computing. In *Communication Systems and Network Technologies (CSNT), 2011 International Conference*, Katra, Jammu, India, 2011.

[4] M. M. Wang, Z. G. Qu, and M. Elhoseny. Quantum secret sharing in noisy environment. In X. Sun, H. C. Chao, X. You, and E. Bertino, Eds. *Cloud Computing and Security. ICCCS 2017*, Springer, Cham. Lecture Notes in Computer Science, vol 10603, 2017, https://doi.org/10.1007/978-3-319-68542-7_9.

[5] Amazon. *Advantages of SaaS-based budgeting, forecasting & reporting* [Online]. Available at: http://aws.amazon.com/ [Accessed January 19, 2012].

[6] D. Neumann et al. Stacee: enhancing storage clouds using edge devices. In *Proceedings of the 1st ACM/IEEE Workshop Autonomic Computing in Economics*, San Francisco, CA, 2010.

[7] Cloud Security. *Assessing the security benefits of cloud computing* [Online]. Available at: http://cloudsecurity.org/blog/2008/07/21/assessing-the-security-benefits-of-cloud-computing.html [Accessed February 16, 2012].

[8] J. Oberheide, E. Cooke, and F. Jahanian. CloudAV: N-vrsion antivirus in the network cloud. In *Proceedings of the 17th USENIX Security Symp*, San Jose, CA, July 2008.

[9] R. Fakoor et al. An integrated cloud-based framework for mobile phone sensing. In *Proceedings of hte ACM SIGCOMM MCC Workshop*, New York, NY, 2012.

[10] D. H. Le Xu, V. Nagarajan, and W.-T. Tsai. Secure web referral services for mobile cloud computing. In *IEEE Int'l. Symposium Mobile Cloud, Computing, and Service Engineering*, Redwood City, 2013.

[11] B. Hay, K. Nance, and M. Bishop. Storm clouds rising: security challenges for IaaS cloud computing. In *System Sciences (HICSS), 2011 44th International Conference on*, Hawaii, 1–7, 2011.

[12] C. Oppenheimer. Which is less expensive: Amazon or self-hosted? Which is less expensive: Amazon or self-hosted?, *Digg Topnews*, February 11, 2012 [Online]. Available at: http://digg.com/newsbar/topnews/which_is_less_expensive_amazo n_or_self_hosted [Accessed February 14, 2012].

[13] E. Toews, B. Satchwill, R. Rankin, J. Shillington, and T. King. An internationally distributed cloud for science: the cloud-enabled space weather platform. In *Proceedings of the 2nd International Workshop on Software Engineering for Cloud Computing*, New York, NY, USA, 1–7, 2011.

[14] W. Dawoud, I. Takouna, and C. Meinel. Infrastructure as service security: challenges and solutions. In *Informatics and Systems (INFOS), The 7th International Conference on*, Cairo, Egypt, 1–8, 2010.

[15] C. Gong, J. Liu, Q. Zhang, H. Chen, and Z. Gong. The characteristics of cloud computing. In *Processing Workshops (ICPPW), 2010 39th International Conference on*, San Diego, CA, 275–279, 2010.

[16] Z. Mahmood. Cloud computing: Characteristics and deployment approaches. In *Computer and Information Technology (CIT), 2011 IEEE 11th International Conference on*, Paphos, Cyprus, 121–126, 2011.

[17] Red Hat Cloud Computing Team. *Red Hat\talking-to-many-clouds-with-the-deltacloud-api*, July 26, 2010.

[18] F. Hu and P. Hu. An optimized strategy for cloud computing architecture. In *2010 3rd IEEE International Conference on Computer Science and Information Technology (ICCSIT)*, Chengdu, China, vol. 9, 374–378, 2010.

[19] J. C. Mudge, P. Chandrasekhar, G. S. Heinson, and S. Thiel. *Evolving inversion methods in geophysics with cloud computing - a case study of an eScience collaboration*, 119–125, 2011.

[20] C. Mudge. *Large scale geoscience workflows in the cloud*, 2011.

[21] W. Lau. *Comparing IAAS and PAAS: a developer's perspective, A Cloudy Place*, January 13, 2012. [Online]. Available at: http://acloudyplace.com/2012/01/comparing-iaas-and-paas-a-developers-perspective/ [Accessed February 25, 2012].

[22] Network World. *5 problems with SaaS security* [Online]. Available at: http://www.networkworld.com/news/2010/092710-software-as-service-security.html?page=4 [Accessed February 16, 2012].

[23] Network World. *Best security questions to ask about SaaS* [Online]. Available at: http://www.networkworld.com/news/2009/031209-saas-security.html [Accessed February 16, 2012].

[24] W. Petruska. *How university data backup is moving online*. Presented at the *Educause Conference*, Anaheim, CA, Oct 15, 2010.

[25] X. Wang, B. Wang, and J. Huang. Cloud computing and its key techniques. In *Computer Science and Automation Engineering (CSAE), 2011 IEEE International Conference on*, Shanghai, China, vol. 2, 404–410, 2011.

[26] J. Martins, J. Pereira, S. M. Fernandes, and J. Cachopo. Towards a simple programming model in cloud computing platforms. In *2011 First International Symposium on Network Cloud Computing and Applications (NCCA)*, Toulouse, France, 83–90, 2011.

[27] The Cloud Foundry Team. *Multi-language, multi-framework, what about Multi-Cloud?* Jan 26, 2012.

[28] google.com, *What is Google App engine? - Google App engine - Google code* [Online]. Available at: http://code.google.com/appengine/docs/whatisgoogleappengine.html [Accessed February 25, 2012].

[29] B. Chun et al.. Clonecloud: elastic eexecution between mobile device and cloud. In *Proceedings of the 6th Conference Computer Systems*, 301–14, 2011.

[30] S. Kosta et al. ThinkAir: dynamic resource allocation and parallel execution in cloud for mobile code offloading. In *Proceedings of the IEEE INFOCOM*, 2012.

[31] A. Covert. *Google Drive, iCloud, Dropbox and more compared: what's best cloud option?*, *Technical Review*, 2012.

[32] A. Abdelaziza, M. Elhoseny, A.S. Salama, and A.M. Riad. A machine learning model for improving healthcare services on cloud computing environment. *Meassurement*, 119:117–128, April 2018, https://doi.org/10.1016/j.measurement.2018.01.022.

[33] A. Abdelaziza, M. Elhoseny, A. S. Salama, and A. M. Riad. Intelligent systems based on cloud computing for healthcare Sservices: a survey. *International Journal of Computational Intelligence Studies*, 6(2/3):157–188, 2017.

CHAPTER 4

Biometrics as a Service

Nitigya Sharma, Prerna Sharma, and Moolchand Sharma

CONTENTS

4.1 INTRODUCTION

The past decade has seen a global surge in the utilization of biometrics technology in many sectors. Not only rich private organizations but also governments have started examining the uses of this technology for security access control, verification, attendance, time-keeping authentication, etc. Biometric national identity systems have been adopted by governments of many countries. An increasing number of enterprises have begun to require rapid implementation of these systems, with hosting on the cloud as a service an efficient solution.

Biometrics maps a person physiologically and behaviorally using imaging, computing, and statistical and mathematical processing of their unique features. For example, in the fingerprint recognition system, an image of the fingerprint is taken, and the many points that make a fingerprint unique are utilized to prepare a matching template which is stored in the database. The input is converted to a comparable data form and compared with templates to match for authentication or identification.

Biometric methods have proven much more efficient, convenient, and secure than traditional authentication methods like PINs and passwords. The major positive of biometrics is the non-shareable and non-forgettable nature of the keys, as they are inheritance-based and only the person himself has access to it. The technology becomes even more efficient when it is combined with multi-factor authentication capabilities.

Biometrics as a Service (BaaS) requires a central data server storing biometric information in a database, fast networking capabilities, and a smooth application to deal with verification/authentication requests from remote devices. The client requires equipment capable of recording a biometric sample. The process records a person's biometric data, associates it with their identity and stores it in the online database. This process is called enrollment. At a later stage, the person can utilize their equipment to request entry to the designated services; the request is sent to the central server, and the matching result is recorded. This step is called authentication.

4.2 CLOUD AND MOBILITY

In the current technologically advanced digital landscape, both consumers and enterprises are enjoying a rapid shift towards the mobile age. With more and more organizations turning towards faster, highly efficient and profitable "on-the-go" approaches, powerful hand-held mobile devices such as cell phones, smart tablets, and netbooks are available to perform day-to-day business transactions for mobile consumers working in a multi-device environment [1]. The increasing use of personal or dedicated employee-held devices has created the need for organizations to establish a "bring-your-own-device" (BYOD) policy [2]. The escalation of data-driven equipment, applications, sensors, and delivery platforms towards the cloud is making communication easier, more coherent and more mobile for organizations and consumers alike. Mobility has been a driving force causing a surge across many industries, mobilizing the workforce and generating enormous benefits in sectors that deal with mobile consumer services. While finance, banking, and insurance companies have been the major players to use portable equipment, healthcare, retail, mass media, and e-commerce have followed suit, developing a more digital profile by including these mobile-centric systems, applications, and services.

With a rapid decline in traditional payment methods such as cash, checks, and demand drafts, emerging digital payment technologies like debit/credit cards, E-wallets, etc. have revolutionized the financial services sector, creating more secure, flexible and highly convenient methods for cashless financial transactions [3]. Mobile and online portals have enabled banks to utilize mobility to grow customer engagement, offering convenience and generating competitive advantage in services such as account servicing, investment portfolios, and transactional engagements [4].

Technology has also revolutionized shopping and retailing, creating a whole new architecture for how people shop and how shops work. The growth of mobile equipment has led to skyrocketing growth in e-commerce and mobile commerce. Retailers have begun to prioritize investment in e-commerce solutions, sidelining old in-store IT systems as they prepare for the predicted increase in their business from online buyers and mobile shoppers [5].

Organizations and service providers are also upgrading their management policies and beginning to adopt new mobile mechanisms for the whole organization, ranging from corporate information systems and customer engagement to purchasing patterns. Mobile equipment greatly increases efficiency and productivity, enhances supply-chain operations, and creates agility for more productive business operations while supporting real-time engagement with consumers, suppliers, and other partners. Case studies and reports show that most of the organizations that are shifting towards mobility and a "virtual enterprise" framework are likely to experience enhanced efficiency and operational advantages in their respective domains.

Many employees have started using their personal mobile devices for banking and e-commerce, accessing sensitive personal information and engaging in payment-related operations. While BYOD provides a large number of advantages in terms of efficiency, productivity, and work flexibility, it is also more vulnerable to cyber-attack due to the lack of security arrangements such as anti-virus and firewalls commonly available on a personal computer [6]. The use of mobile devices for simple tasks and functions intensifies the need to safeguard data, equipment, and networking against loss, compromise, intrusion—anything that poses a risk to the organization and individuals' private information. The safeguarding of their data, whether stored on workers' or corporately owned equipment, is critical for any organization.

While many enterprises seek to take advantage of the mixture of policies and techniques offered by BYOD, the risk to consumers is increased, with web-based intrusions by malicious software that resides on unprotected sites and devices making it easy either to access consumers' personal data or to breach the organization's security[2]. With the increased popularity of BYOD within organizations, cyberattacks are likely to become more severe, frequent, sophisticated, and capable of harming a part of the bigger picture, i.e., a specific section or department of the enterprise.

4.3 COMPLEX THREATS AND THE CYBERSECURITY LANDSCAPE

BYOD offers numerous benefits to enterprises: greater efficiency and flexibility, cost saving in information systems, increased job satisfaction for employees, and a much faster system with reduced response time. However, letting the workers use their own equipment to work and access the organization's data stores, which are very important for the business, could also intensify security threats and risks. The development of mobile and online banking, as well as the increasing trend towards e-payments, are encouraging e-commerce, service providers, and the banking sector to seek strong and advanced security solutions to safeguard their services against criminals targeting their consumers and compromising their information.

4.3.1 Cloud-Enabled BYOD

An architecture based on cloud services has the potential to offer convenient, flexible, cost-effective, and secure solutions. New technologies in cloud-enabled applications assist developing operating environments and business workflow, with the central aim of enhancing business efficiency and productivity [2]. Working in a location-fluid environment provides numerous opportunities for innovation and advancement, with mobility providing the unique advantages of routes and location. Over time, more and more enterprises will be seeking to use cloud solutions to store their critical data and applications on public cloud platforms. Organizations increasingly using cloud-enabled applications and mobile equipment for critical business processes and saving their data and information in public cloud services are at increased risk of data breaches, fraud, and many different malicious activities if there is no proper system to counter cyberattacks [6]. As time will proceed, more and more enterprises will seek to utilize cloud solutions to store their critical data and applications on public cloud platforms.

4.3.2 Virtualization for BYOD

Virtualization in the desktop enables workers to access the workplace and do their work anywhere and anytime across a wide collection of devices ranging from laptops and desktops through to tablets and smartphones. The virtualization of the desktop significantly reduces the costs of physical hardware, software, and maintenance. It also empowers the IT department to manage and support these devices with the required operating systems and software centrally simply by ensuring they are connected over a network. It is also expected to develop productivity and the continuity of the business. Virtualization techniques include hosted virtual desktop (HVD), desktop as a service (DaaS), virtual desktop infrastructure (VDI), and other server-based computing models.

All the security challenges facing a full-fledged physical desktop also affect virtual desktops. Access security is the most common threat to a virtual machine. Anyone who accesses the user's user name and linked password can illegally breach the virtual machine.

4.3.3 Emerging Mobile Business Workforce

New equipment and the potential to use any service anywhere and anytime has encouraged enterprises to use them for business processes and streamlining workflows. A significant number of organizations have already started employing BYOD solutions. Collaborations and associations, communication, and social technologies have become crucial tools for arrangements and task scheduling. Mobile equipment is revolutionizing the organizational workflow for top professionals with an increasing number of real-time business-centric applications for business service delivery models. The high level of IT customization resulting from BYOD policies not only empowers the workforce but also plays a significant role in reshaping the look and feel of an enterprise's applications. Shadow IT is emerging both as a necessity and an open loop for security threats, with a tech-savvy workforce continuously looking for solutions to any problems with the business line. Real security concerns start to arise when any kind of unsupported or unwanted hardware/software is used that does not meet the requirements set up by the organization for security, control, documentation, and reliability, leading to a higher risk of uncontrolled, unmonitored, and unofficial flows of data. It is generally the case that when employees are allowed to self-administer the endpoint applications of an enterprise, its rules and security policies become a complex and hideous task. This kind of vulnerability within the IT department leading to shadow IT should be carefully identified by the organization's decision makers.

4.3.4 The Financial and Banking Sector

The rise in networking technologies leading to the development of online and mobile services has done wonders to reshape the banking sector. Banks have become more efficient and convenient, faster and greener. Banks look to technology to accurately identify potential customers, provide them with flexible, secure, easy-to-use banking service methods, enrich core customer-relations systems, and increase customer satisfaction and retention. Banks and e-wallets have a number of methods by which customers can access their accounts using online access via a web browser or a dedicated mobile application [4]. However, in line with the increased availability of these services, an increase in fraudulent transactions, identity theft, and data breaches is also observed. Financial institutions are therefore continuously working to improve the security architecture of their business processes beyond traditional authentication based on user name and registered password.

4.3.4.1 Challenges

In the digital age, more than half of the world's population has access to some kind of mobile device, making online platforms a very hot target for cybercriminals. Bank accounts have become more vulnerable, leading to trust issues and distressed customers. While online and mobile banking solutions are fast, convenient, and easy, they are no match for the customer satisfaction of a face-to-face transaction, and the trust built up by the traditional brick-and-mortar banking system is somewhat tarnished by these systems. Moreover, there is a very thin line between security and inconvenience. The heavy reliance of the banking system on PINs and passwords sometimes renders their effort to provide a seamless service an inconvenient process [4]. All this has motivated banks to research more robust and reliable security measures that can efficiently and accurately detect frauds and prevent criminal activities such as data breaches and hacking of customer accounts.

4.3.5 E-Commerce

Mobile shopping and electronic commerce, or e-commerce, have developed massively in the past two decades. A shift in customer buying behavior, the easy availability of credit and finance, the large number of online shopping sites and mobile commerce applications, fast and seamless payment methods, together with widespread access to portable equipment and applications, have led to the rise of this sector. E-commerce has started to capture a large market share, increasing steeply year by year. It has changed traditional ways for companies to do business and the buying preferences of customers. Product showcasing problems have been eliminated, with a huge number of products easily showcased online and most retailers having access to a larger than ever customer base and supplier community, which empowers them to expand their market out of its preset boundaries to the international market [5]. For customers, buying has never been easier. Anything can be bought online with a live experience 24/7; user reviews help potential customers to make faster and better choices. In addition, the availability of numerous convenient payment options provides a marvelous shopping experience.

4.3.5.1 Challenges

In the mobile and online commerce landscape, fraudulent transactions, payment credential-related data breaches, and identity theft-related crimes have been the most commonly applied cybercrime techniques. E-commerce is one of the most affected platforms and has faced the largest share of attacks; this trend is expected to continue for some years due to the large customer domain providing more and easier targets for attackers. The main challenge for e-commerce is that no face-to-face interactions are involved: the customer and their device are remote, so the seller cannot know who is actually making the transaction [5]. Authentication measures based on PIN and password are utilized, but these become void where PINs or passwords are stolen or used without permission, leading to compromised data, loss of valuable customers, and an increment in the industry's operating costs. E-commerce is highly affected by these challenges and there is a continuous need for ways to set up transactional security measures that can increase security for a wide variety of transactions.

Organizations in this field have started recognizing the continuously evolving threat to BYOD policies in the online environment and the resulting increase in the number of cyberattacks. Drastic measures are required for the protection of private and sensitive 28.9 million in 2013. With such a rise in frauds, cyberattacks, and data breaches, customers have got personal or organizational data and strengthen national security. A report by PWC states that in 2014 the number of security infiltrations increased to 42.8 million, an enormous increase of 48%. Customers feeling a sense of insecurity are seeking relatively safer ways of making purchases on online and mobile platforms. This is widely acknowledged by enterprises, which in consequence are moving to newer and more advanced security practices such as I&AM (identity and access management) systems.

The following sections explore the challenges and complex IT security issues faced by I&AM.

4.3.5.2 Variety of Devices and Operating Software

The response of mobile applications and businesses to advances in technology have resulted in an increase in availability of different equipment and services requiring different support and management, which has in turn led to more complex IT management of customer services, devices, and access.

4.3.5.3 Identity and Access Management Difficulties in the Workplace

Promotions and transfers make for frequent role changes for workers in organizations. This often makes it difficult to keep track of employees' entitlement to information access, such that their access rights do not correctly reflect their current role, and they may retain access rights relating to a previous role. To ensure that employees are granted the correct, controlled access rights based on their roles in the organization, managers must take the time to map IT system data with HR data.

- Due to badly organized or faulty data storage schemes, identity-related employee information is often duplicated on multiple systems, leading to inconsistencies with data updates that are stored discretely and managed independently.

- Traditionally, in IT organizations a dedicated person performs administrative tasks such as distributing access control according to the employee's profile. However, issues may arise between a monitoring system and a human administrator in an I&AM environment, such as:

  - *Account cloning*: workers performing similar roles for an organization may be granted the same access level.

  - *Orphan account*: accounts previously assigned to a person who either left or was promoted. If not properly managed, these orphan accounts appear in the system without a legitimate owner.

  - *Mixed access rights*: employees are provided with access rights to places where they are supposed to work, but often a combination of access rights can lead to security risks.

I&AM systems are designed to perform various tasks such as program-based system deployments, management of user identity, and entitlement management, efficiently and at a moderate cost. System design also incorporates security. As the number of people accessing sensitive data in the course of their job increases, the organization's IT function is required to periodically update critical data, security parameters, and other technologies to restrict losses from cyberattacks. As more and more processes emerge in the digital landscape, mitigation of security risks and increased protection of data becomes essential.

4.4 EMERGENCE OF BIOMETRICS

> The word Biometrics comes from the Greek words "bios" (life) and "metrics" (measure). Strictly speaking, it refers to science involving the statistical analysis of biological characteristics. Thus, we should refer to biometric recognition of people, as those security applications that analyze human characteristics for identity verification or identification. [7]

Biometric identifiers can be broadly divided into two distinct categories: behavioral characteristics and physiological characteristics. This technique can efficiently and accurately identify a human using distinct physiological characteristics, for example through facial recognition, fingerprint recognition, retina, palm vein, and iris recognition. Behavioral biometrics are designed to analyze uniquely identifiable and measurable human characteristics such as gait, voice, and typing rhythm. The input for a biometric system is usually processed by a pre-defined algorithm based on the biometric type, which converts the input to data points. These data points are passed through a scanner where they are matched with the pre-determined sample, leading to user authentication.

Data security has become an important aspect of information technology, with customers and organizations incorporating the use of electronic equipment in almost all dimensions of their life. A secure digital environment wave is largely being carried by the rising demand for smart devices that is engulfing both consumer and enterprise markets [8]. Enhanced authentication measures help us to take on new technology tools that offer a great advantage over traditional access control mechanisms.

4.4.1 Ubiquitous Biometric Technology

Many varied domains have started utilizing biometric solutions to make their authentication procedures increasingly secure. Much more convenient than other measures, its growth has been accelerated by the following benefits:

- Unique – For a particular kind of biometrics, no two people can have a similar data point formation.

- Inimitable – Biometric input terminals use living identities for validation.

- Unshareable – No one can share their own biometric characteristics, leading to protection against theft or duplication.

These features are exclusive to biometrics technology, which is applicable to both intellectual property and organization or individual data. The technology provides an enhanced level of security compared to traditional authentication methods like user IDs, PINs, passwords and other access management technologies [7]. The growth in biometrics technology can be predicted by watching the rapid growth of the fingerprint recognition market over the past couple of years, with the technology now included in most kinds of smartphones and other devices.

Increasing awareness of the usefulness of biometrics for authentication and identification will secure a stronghold for this technology in the global commercial market and see its adoption for a large range of user applications such as healthcare, banking, border control, and secured area access authentication.

In mobile commerce and online shopping industries, new methods are being enthusiastically adopted to provide customers with important product and price information, together with related reviews. An increasing proportion of consumers have started opting for online shopping with all the advantages of fast selection and easy delivery. Retailers face a challenge, with more and more customers moving to mobile payment options, revolutionizing the consumer market. This has led to an urgent need to deploy efficient and fast online services that can manage all kinds of online payment transactions. The inclusion of biometrics in electronic commerce has produced a safe, secure, and convenient payment method, making for a slick and smooth customer experience. A leading e-commerce retailer in the United States describes the benefits of BaaS for the global market:

> One of the biggest concerns consumers today have is the risk of fraud when shopping online. Savvy cybercriminals are a threat to e-commerce and the online shopping industry diminishing business opportunities in this sector. The

top challenge in retail today is to secure the Point of Sale (POS) and Omni channel retailing processes where unified security measures will provide true continuity of customer experience. The high-tech biometrics authentication system is the future of secure sales. [9]

4.5 TAKING BIOMETRICS TO THE CLOUD

Cloud technology represents the next upgrade, one that supports the all-round market demands of the current digital economy. Services in the cloud can be accessed anytime from anywhere in the world. Some of the advantages provided by cloud technologies that are being heavily utilized by organizations of all sizes are[10]:

- *Network-centric*: conveniently available on any type of public network or equipment.
- *Controllable resources*: can be shared according to customer need and economic situation and can be increased and decreased based on demand and cost.
- *Agility and elasticity*: services and resources can be deployed across varying infrastructures.
- *Data sovereignty*: one of the main traits of BaaS, this is a very helpful tool for many organizations.
- *Speedy infrastructure deployment.*
- *Biometrics on demand*
- *Economically affordable Need-based rapid scalability*
- *Non-Redundant*
- *Convenient customization*

Table 4.1 compares cloud-based with on-device solutions.

TABLE 4.1   Cloud-Based vs On-Device Solutions

Cloud-Based Solutions	On-Device Solutions
• Automation provides cost savings	• Costly to develop endpoint-compatible solutions
• Central access across all services and devices	
• Automatic data back-up	• No central management capability Individual handling is mandatory
• Seamlessly integrated with internal and external services	• Data back-up has to be done by end user
• Threat-prevention measures in place	• Integration not scalable
	• Poor defense against threats
• Multidimensional security solutions covering all endpoints	• Security solutions are individualistic

4.5.1 Biometrics Meet Banking

Since the adoption of digital techniques for banking processes, the benefits provided by automated cloud computing services have reduced costs of operational and administrative tasks. Biometrics have played a crucial role in providing high levels of security and cutting-edge banking solutions for bank customers [4]. Biometric techniques genuinely identify an individual, in contrast to traditional authentication systems based on a set piece of information only. An inherent characteristic of biometric information is that it cannot be shared or forgotten. Major banks are utilizing biometric methods for a wide range of services such as ATM withdrawals, fingerprint-based sign-in, or face/voice verification for mobile applications. Banks can set up their classified information on a central cloud server and utilize a third-party biometric authentication system for verification, making it highly scalable and appropriate to meet ever-rising customer security demands.

A three-step banking system with biometrics can be set up as follows:

Step 1:

- Log in to the bank's website hosted online via a PC
- Sign in to the bank's mobile app using a compatible mobile device

Step 2:

- Verify the banking transaction from the bank's website
- Verify the transaction via the bank's mobile app

Step 3:

- Verify ATM transactions
- Authorize/decline unusual transactions and prevent fraudulent activity
- Authorize payments and services on a periodic basis

4.5.2 Secure Payments

Online shopping and M-commerce enterprises are being challenged by the rising number of digital customers to provide secure convenient digital payment methods. The required high levels of security and protection against theft, fraud and other forms of cybercrime can be rapidly achieved with the help of biometrics in the cloud, the technology that is currently favored. Under this architecture the identity of the person can be actually verified.

FIDO (Fast Identity Online) [11] is a standard universal password replacement mechanism that uses two-factor authentication. The FIDO standards comprise multi-factor authentication (MFA), near-field communication (NFC), USB tokens, and numerous other authentication options to mimic traditional password systems.

Backed by industry leaders including Microsoft, PayPal, and Lenovo, FIDO uses two methods to deal with authentication needs: (i) a universal authentication framework (UAF) which uses a public key for logging on and supports user equipment with PINs or fingerprints; and (ii) a universal 2nd factor (U2F) which counter-authenticates the user with a PIN or biometric passcode. A further advantage of two-factor authentication is standardization of systems and increasing interoperability, leading to a significant cost reduction for payment industries.

4.6 BIOMETRICS AS A SERVICE APPROACHES

From commercial I&AM systems already in use to developing biometric systems, multi-modal biometric technology has now advanced to use various kinds of distinct sensors to extract a variety of information from a single marker. This includes multiple images for fingerprints or iris scans, leading to reliable recognition and stronger security, and greatly increasing accuracy, as biometric information can also be detected correctly from worn fingerprints or damaged irises [12].

BaaS architecture looks somewhat complex but is in fact remarkably simple [13]. It simplifies the deployment of any kind of biometric service over the cloud easily and conveniently. The architecture (illustrated in Figure 4.1) consists of the following entities:

1. *Company server*: stores and manages all the company data on the central cloud. All data and service needs of the company are fulfilled by this server.

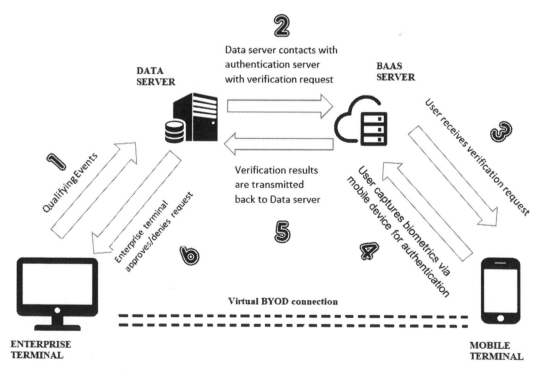

FIGURE 4.1   BaaS architecture.

II. *Authentication server*: the middle server that performs the authentication and security tasks. This server is assigned to authenticate different users looking for access to the system.

III. *BYOD client device*: end terminals used by the workers or clients to access services.

IV. *Server biometric application*: services requests over the cloud for biometric processes.

V. *Client biometric application*: application installed on the device that integrates the biometric capabilities of the equipment with it and provides the authentication service on it.

The company stores all the data and information required by different layers of employees on a server hosted in the cloud. This server does not directly access the secure authentication server that manages and authenticates access by users in the company's server. The server application on an authentication server manages calls made by BYOD client devices and fulfills their requests [14]. The BYOD client device houses the biometric client application through which access requests are made. The architecture is capable of:

- Creating a biometric profile and enrolling users.
- Saving biometric samples on the cloud.
- Verifying users by comparing input samples with stored biometric data.
- Setting up notification and access communications (Table 4.2).

4.6.1 Market Challenges with Biometrics

With more and more enterprises adopting mobile processes, companies' stores of both personal and customer data are set to increase exponentially. Large numbers of people are readily sharing their identities and data with organizations that request it for analysis and the valuable insights it provides [7]. Privacy concerns are very significant in biometric-based identification technologies. Though biometrics solve many of the problems associated with traditional password and paper-based methods, it has given rise to unique concerns related to biometric security [8]. For example, it is quite possible for a person's facial features to be captured and stored by a facial recognition system without their permission and utilized for fake authentication. Likewise, in fingerprint detection systems, it appears that the system can be fooled by taking fingerprints from hard surfaces that people have touched. Another privacy challenge is the unauthorized use of collected biometric data by organizations for other purposes. Lastly, analysis of biometric data can reveal information such as health, occupation, and socioeconomic standing, violating the anonymity of individuals.

Biometrics is very device oriented: it depends on the capability of the client device hardware. All biometric technologies require robust equipment that can conveniently take biometric samples and store them on the device for further use. Another challenge is the diminishing need for standard solutions in a multi-party environment, with standardized

TABLE 4.2   Examples of the Solutions Provided by BaaS in the Fields Of Healthcare and Banking.

Healthcare Challenges	BaaS Solution
• Secure storage and communication to safeguard crucial patient-related data	• Hospital staff can access central patient information at their own desk.
• Hospitals are abandoning traditional password-based unsecured authentication methods	• Patient record sharing will become more convenient
• Safer and more convenient patient verification processes required	• Eliminates chances of patient identification error to a great extent
• Fraud detection and anti-duplication systems required	
Banking Challenges	**BaaS Solution**
• Banking sector needs very secure and seamless identification process to detect data breaches and eliminate fraud	• Mobile application or online browser can be used to verify transaction after secure log-in
• Banking requires security in riskier applications services such as mobile banking, branch banking, ATMs, online transactions, and data network secure sign-ins	• ATM, online banking, and E-payment authentication can be done more securely
• Improved customer trust in mobile banking and online transaction mechanisms	• Biometrics-powered fraud prevention system makes banking safe and secure

methods unable to meet the needs of users requiring a variety of methods to deal with a particular problem.

4.7 FUTURE CAPABILITIES

In the government sector, agencies responsible for issues such as national identification, border patrol, and internal law enforcement have been among the earliest to utilize biometrics. With the increase in mass access to mobile devices, it can be predicted that the use of biometric technologies for authentication purposes and online transactions across various industries will steeply increase in the coming years.

Data-sensitive industries like banking have started investing in the new developing biometric technology. Large-scale investments are already occurring in biometric capabilities, which are expected to double in the near future. A large number of insurance companies are moving towards biometric solutions to handle their complex ecosystem, and it is expected that a significant proportion will take up BYOD policies for their workflow. Increased investment can be expected from a financial sector seeking to acquire efficient mobility solutions and enterprise applications. Biometric verification and authentication tasks to manage customer relations and security will lead to this technology becoming an important tool in the future. A leading Australian insurance company states:

> The insurance business these days is geared towards providing on the door customer experience and services. Captive and geographically dispersed agents and brokers are mobilized, improving sales rates, customer visibility, and

productivity by embracing BYOD. Biometrics for verification offers convenience for agents and customers for high levels of security access for insurance policies and transactions. [3]

The application of biometrics technology for identification purposes will see growth in numerous sectors, including information and communication, healthcare, cloud storage, smartphones, and banking. Once Apple had become the first company to introduce fingerprint technology in smart devices, fingerprint recognition became a must-have in subsequently developed equipment. In the race to advance, other recognition biometrics will follow in the smartphone industry.

REFERENCES

[1] P. M. Rajendra, R. Lakshman, and V. Bapuji. Cloud computing: research issues and implications. *International Journal of Cloud Computing and Services Science*, 2:134–140, 2013. 10.11591/closer.v2i2.1963.

[2] E. Meisam, V. Maryam, H. Habibah, T. Noorita, and S. Ezril. BYOD: current state and security challenges. In *ISCAIE 2014 - 2014 IEEE Symposium on Computer Applications and Industrial Electronics*, Penang, 2014, 10.1109/ISCAIE.2014.7010235.

[3] Digital Insurance News. August 1, 2017 [Online]. Available at: https://www.munichre.com/site/marclife-mobile/get/documents_E2104338863/marclife/assset.marclife/Documents/Publications/DigitalInsurance_Biometrics_9.18.17.pdf

[4] Z. Zahoor, M. Ud-din, and K. Sunami. Challenges in privacy and security in banking sector and related countermeasures. *International Journal of Computer Applications*, 144(3):24–35, June 2016.

[5] S. Kunal, S. Amarjeet, and P. Sharma. SMEs and cybersecurity threats in E-Commerce. *Edpacs*, 39:1–49, 2009, 10.1080/07366980903132740.

[6] O. Morufu, T. Abdullah, M. Mahmod, and R. A. Azizol. A review of bring your own device on security issues. *SAGE Open*, 5, 2015, 10.1177/2158244015580372.

[7] F. Z. Marcos. Biometric security technology. *Aerospace and Electronic Systems Magazine, IEEE*, 21:15–26, 2006, 10.1109/MAES.2006.1662038.

[8] B. Debnath, R. Rahul, A. Farkhod, and M. Choi. Biometric authentication: a review. *International Journal of u- and e-Service, Science, and Technology*, 2, 2009.

[9] Frost & Sullivan. *Cloud-based identity and authentication: Biometrics-as-a-service, a White Paper by Frost & Sullivan in collaboration with Fujitsu.*

[10] T. Yu and Y. Zhu. Research on cloud computing and security. In *2012 11th International Symposium on Distributed Computing and Applications to Business, Engineering & Science*, Guilin, 314–316, 2012, 10.1109/DCABES.2012.28.

[11] FIDO Alliance White Paper: FIDO Authentication and the General Data Protection Regulation (GDPR), May 2018.

[12] M. Ali and A. Gaikwad. Multimodal biometrics enhancement recognition system based on fusion of fingerprint and PalmPrint: a review. *Global Journal of Computer Science and Technology*, 16:13–26, 2016.

[13] A. Castiglione, K. K. Choo, M. Nappi, and F. Narducci. Biometrics in the cloud: challenges and research opportunities. *IEEE Cloud Computing*, 4:12–17, 2017, 10.1109/MCC.2017.3791012.

[14] H. Vallabhu and R.V.S. Satyanarayana. Biometric authentication as a service on cloud: novel solution. *International Journal of Soft Computing and Engineering (IJSCE)*, 2(4):163–165, September 2012, ISSN: 2231–2307.

CHAPTER 5

The Role of Intelligent Grid Technology in Cloud Computing

Mansi Jain, Parmita Kain, Deepak Gupta, and Joel J. P. C. Rodrigues

CONTENTS

5.1 INTRODUCTION

Cloud computing has evolved to enable remote servers on the internet to store and manage data using a database, and to process data instead of using a local server. Hardware and software models have been brought together to provide utility solutions for various programming problems.

In the early days of the internet, expensive specific servers had to be purchased, and troubleshooting problems was in conflict with business goals. With the advent of cloud computing these problems have been solved. Cloud providers manage and hold our data and servers, and we no longer need to look after the infrastructure. It is a win–win situation. The cloud has higher processing power and faster access, as well as providing authorized access to the various networks in the grid.

The idea of grid computing was first proposed by Ian Foster and Steve Tuecke of the University of Chicago, and Carl Kesselman of the University of Southern California's Information Sciences Institute. They developed a standard toolkit to store, process and compute data. The concept mainly targeted scientific and research projects in which large computing units are collaborating. Grid technology deals with interoperability between providers. It does not require any fundamental business model.

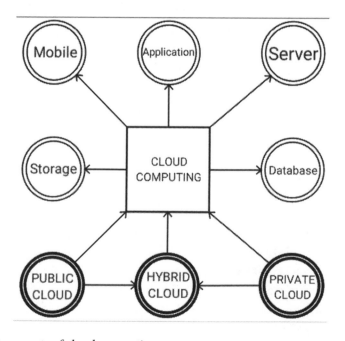

FIGURE 5.1 Main aspects of cloud computing.

In the business model known as utility computing, a provider holds and manages the computing resources and techniques on a pay-per-use basis. This is suitable for software (SaaS), platform (PaaS) and infrastructure (IaaS) applications.

Though cloud computing, grid computing and utility computing all control and exploit shared resources, their architecture, type of institution and functionality differ. In this chapter, we examine the various aspects of grid computing and their role in cloud computing (Figure 5.1).

5.2 FUNDAMENTALS OF GRID AND CLOUD COMPUTING

5.2.1 Grid Computing

Grid computing connects resources like processors, storage, data, applications and so on, and provides access to them. PCs that are connected together to gather information form single computable intensive applications. In general, grid computing requires at least one computer, i.e., a server. This server handles all the administrative work for the system. A middleware is also required. The workload will be divided among different systems working towards a common goal. The user sends a request for application execution and the resource broker chooses a domain with appropriate resources [1–3].

For dependable, compatible, cost-effective retrieval of computational capabilities, a computational grid is required. The grid can be thought of as a distributed system with a noninteractive workload that involves a large number of files. The basic requirements for grid computing are the client/user, applications and data (see Figure 5.2).

Some important advantages of grid computing are:

- It uses underutilized resources.

- The grid uses parallel CPU capacity.

- Through virtual resources and virtual organization, the grid is used for collaboration.

- The grid enables free access to resources.

5.2.1.1 Classification of Grid Operations

The data grid is a system for handling, modifying and transferring large amounts of distributed data sets. Here, data can be located, whether in single or multiple sites depending

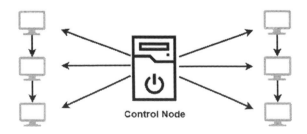

FIGURE 5.2   How grid computing works.

upon the requests made, through middleware applications and services that pull the data requests from various administrative domains.

The CPU scavenging grid is a cyclical movement of projects from one PC to another according to user requirements. It is a form of shared computing that takes unused resources from a wide network of different participants in an organization or across a wider space.

Grid architecture is not a single-step process. It is a globally defined forum protocol that includes the following:

- A secure grid infrastructure.

- Appropriate services to be monitored.

- Allocation of grid resources and management of protocol.

- Global access to secondary storage and grid FTP.

- Proper storage capacity for computing resources.

5.2.1.2 Benefits of Grid Computing

- Expensive SMP servers not required for smaller applications.

- More appropriate and efficient use of resources.

- Minimum risk of failure due to grid environment's modular structure.

- Easy to manage all policies.

- Easy upgrading without scheduling downtime.

- Equal preference, hence equal speed, given to all jobs.

5.2.1.3 Disadvantages of Grid Computing

- Memory problems.

- Rapid connection between various computing resources needed.

- Less flexibility.

- Political challenges associated with resource sharing [2, 3, 18].

5.3 CLOUD COMPUTING

Cloud computing means the use of remote servers on the internet to store, manage and process data other than on personal computers or laptops. Servers can be rented on an hourly basis, since it is no longer necessary to buy expensive servers for data computing.

Users are able to focus on their goals, free from the responsibility of managing servers. The stability factor in cloud computing comes into play at times of peak traffic, with resources scaled according to the traffic. There are two aspects of the cloud in an internet data centre (IDC): cloud services and cloud computing. Cloud services are any service that can be delivered to the user in real time via the internet. Cloud computing is the distribution of products, services and solutions in real time. Examples of cloud computing services are Amazon EC2, Google App Engine and Apple iCloud.

There are two types of cloud models: service models (SaaS , PaaS, IaaS); and deployment models (private, public, community and hybrid).

Cloud providers use the "pay-as-you-go" model, which, somewhat unexpectedly, controls the price.

The reasons for the growth of cloud computing are:

- Accessibility of high-capacity networks

- Computers available at cheaper rates

- Storage devices

- Widespread promotion of hardware virtualization, service-oriented architecture, autonomic and utility computing.

The goal of cloud computing is to provide clients with all the benefits of technologies without them having to acquire detailed knowledge of each of them. The cloud sets cost-efficiency goals and the client focuses on their core business. Cloud computing uses virtualization technology, enabling one or more virtual devices to be separated.

5.3.1 Types of Cloud

Four types of cloud can be identified [15], according to business needs (Figure 5.3):

> *Private Cloud:* In a private cloud, providers belong to one corporation [1, 2, 4]. The computing stock is used by a specific organization, mainly for interactions within the business.

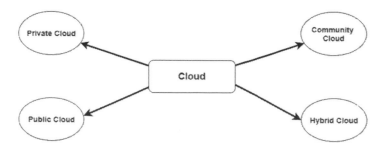

FIGURE 5.3 Types of cloud.

Community Cloud: In a community cloud, computing stock is available for both the community and organizations.

Public Cloud: Mainly used for business-to-consumer (B2C) interactions. The government, business or academic organization owns, governs and operates the computing stock.

Hybrid Cloud: Can be used for both B2C and business-to-business (B2B) interactions. Here, computing stock is held together by different clouds.

5.4 MODELS OF GRID AND CLOUD COMPUTING

The interaction process performs all the tasks in a distributed system, and interactions can be either synchronous or asynchronous. Interaction models are affected by communication performance. It is also impossible to maintain a consistent global time.

Furthermore, in the grid, users of consumer organizations use different resources from more than one resource provider. It is important for consumers to know about interaction models and protocols. The Open Grid Forum (OGF) collects the required standards information, including Job Submission Description Language (JSDL) and Basic Execution Service (BES).

In the case of cloud computing, a user may wish to interact with more than one provider while remaining anonymous. Hence, cloud consumers may order protocols and standardized work, as shown in Figure 5.4 [1, 2, 5].

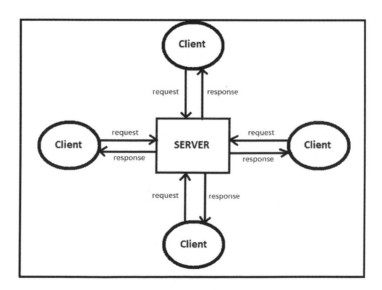

FIGURE 5.4   Client–server relationships.

5.4.1 Distributed Computing

The two areas of work in a grid distributed-system project are:

1. Administration and managing the collection of interoperable resources

2. Working with different middleware.

Available middleware includes gLite, Unicorn, etc. Globus middleware is a de-facto standard. A de-facto standard is one that has achieved a dominant position by public acceptance.

The grid contains a collection of computing resources that may be under different administrative domains, such as universities, but inter-operate transparently to form virtual organizations. The cloud has similar working requirements for supporting complex services, providing users with services on different levels of support, i.e. application, platform and infrastructure. The commercial clouds work in the same way as regards resources and compete on the basis of power and communication purpose [1, 2, 6].

5.5 CLOUD COMPUTING AND GRID COMPUTING COMPARED

Table 5.1 compares the features and benefits of cloud and grid computing.

TABLE 5.1   Comparison of Cloud and Grid Computing

Basis	Cloud Computing	Grid Computing
Definition	A new class of computing with reference to network technology Networked and integrated software and hardware	Divided architecture where networking of computers is used to resolve all problems
Types	Public clouds Private clouds Community clouds Hybrid clouds	Distributed computing systems Distributed information systems Distributed pervasive systems
Goals	Reduction of costs and increased returns. Increased scalability, availability and reliability.	Mainly network focused, therefore has large-scale goals. Resource sharing, uniform and reliable access to data, storage capacity and computation power. Focuses on using a computer as a utility.
Advantages	Can store large amounts of data safely, and data can be accessed at any time Can be easily accessed via internet connection Provides best performance to keep data secure Runs on the latest network Automatic software updates Fast back- up and restoration services	Uses idle energy of the computer effectively Saves money on huge projects Divides work between different computers More reliable, as even if one computer stops, others keep working Space saving
User management	Entire cloud is managed by a centralized system or can be passed on to any third party	Decentralized management Virtual organization-based management

5.6 MODEL

Despite their similarities, approaches to grid and cloud computing system differ on the basis of:

1. Software

2. Infrastructure

3. Platform

4. Application

The basic architecture is shown in Figure 5.5.

5.6.1 Software

Software as a Service (SaaS) in cloud computing uses servers in the cloud to store databases, most commonly used by sales forces. Zoho Office and Google Apps are examples of companies providing this service.

5.6.2 Infrastructure

Infrastructure as a Service (IaaS) is the backend hardware element. Examples include Amazon EC2 or Open Nebula. It is the lowest or most basic layer of an organization's cloud architecture. Server equipment in the cloud makes infrastructure such as the full OS and firewall accessible to other users through APIs or command-line tools, in a similar way to a PBX or telephone system. The benefit to companies is lower IT staffing requirements [16].

There are various steps required for accessibility of the cloud environment.

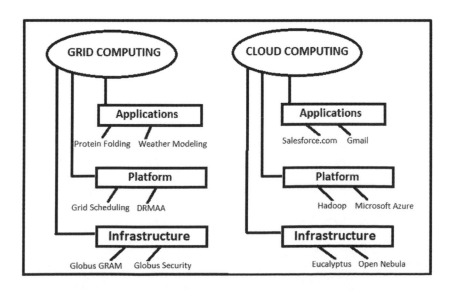

FIGURE 5.5   Basic architecture of grid and cloud computing.

- Infrastructure security measures are taken by cloud vendors such as AWS, Amazon, Microsoft, etc. Companies require a valid certificate issued by an authority, such as Telecom Regulatory Authority of India, and users must be authorized to use the system. Grid systems are managed by the Community Authorization System (CAS). In Amazon or Eucalyptus, web forms are used for fresh arrivals.

- After authentication, an intermediate source must be contacted to approve the request. Most web service requests are sent using API or available tools. In the same way, communication occurs between the scripts and the grid.

- The user then needs to designate the type of action or task to be accomplished on the destination resources. Grid technology supports resource specification languages (RSLs), which define a common path to describe resources and job submission description languages (JSDLs), which describe the computational task submissions.

- Once requests are received, job scheduling and task management are carried out. In grid computing the Globus resource allocation manager (GRAM) enables location, control, job monitoring and cancellation. In cloud computing, RCSTD is used to improve efficiency and balance various tasks. Similarly, SFBAJG is used to improve the communication dependencies and computational capabilities of various resources. In grid computing, scheduling may be homogeneous or heterogeneous, depending upon the type of task or job. For example, the Round Robin scheduling technique is used by Eucalyptus to provide a fixed time interval for access. AWS Batch is used in Amazon to run jobs.

- The final stage is management and supervision of various jobs and resource usage. In grid computing, GRAM and MDS (Monitoring and Discovery Service) are used to ascertain the status of submitted jobs, provide adaptation capabilities for heterogeneous resources and make data available for enquiries. Amazon and other major providers use high-level monitoring techniques.

5.6.3 Platform

Platform as a Service (PaaS) is the web base for the evolution of new or the latest technology. There is no need to install PHP or any updates; the only requirement is to collect or write the code and put it into a single base to run the web application, checking whether the web application formed is applicable to that platform or not. An additional service is thus created without reference to the physical resources required, which are built upon the infrastructure. All software developers, web developers and businesses benefit from PaaS. Examples include Microsoft Azure and Google App Engine. There are two aspects of platform-level solutions:

a) Creation and separation from the physical layer. Below the underlying infrastructure is a lower platform layer for interaction, and this allows the programmer to develop new software, depending upon the network configuration.

b) API service support. In PaaS, developers are mainly focused on building the application through testing. API helps to make the correct choice of application depending upon the features required, allowing the user to develop a new application in accordance with PaaS solutions. There are various aspects of PaaS, including the following:

- Libraries in PaaS provide a database middleware solution, offering additional security for database administrators. Grid middleware allows the user to submit program requests for execution and computation to their grid and to create interoperability between two or more networked computers. Apprenda is one of the companies known for its excellence in database middleware.

- The simple API for grid applications (SAGA) works as an interface and language connection that focuses on handling and sending various jobs.

- MPICH is an API used for connecting the interface with the message-passing interface (MPI).

- A programming model named Grid Superscalar enables applications to run on the grid depending upon their data dependencies as well as writing code on a particular function. It reveals not only dependencies but also the location of the task.

- Another programming paradigm is SWIFT, which builds and defines computations and data dependencies for the efficient running of a large number of jobs.

Though the terms cloud PaaS paradigm and cloud computing are used in the same context, the major difference between them is that the cloud paradigm is used as a business model that provides a new form of computing at cheaper rates, which had previously been thought to be impossible. Therefore, PaaS solutions are designed in accordance with the infrastructure, whereas in grid computing, the work and execution are done manually.

5.6.4 Applications

There is no difference between applications on grid and cloud computing as regards performance and storage, but computations are carried out on the basis of the available APIs and resources [15, 16].

Most grid applications are for scientific software. Technologies that use grid and cloud computing are identified at different levels of acquisition, depending upon the following causes:

(i) Lack of business opportunities in grids

(ii) The complexity of grid tools

(iii) Accordance with the target software

Cloud and grid computing have applications in a variety of sectors:

1. *Education*: cloud computing is used in smart classes to make digital education available to all students, increasing learning opportunities.

2. *Industry*: it solves various technical and business problems that can occur during data illustration, reduces costs and helps manage records more appropriately.

3. *Healthcare*: helps with patient record keeping and only allows authorized access.

4. *Banking*: Does not provide security facilities but assists with communication services and core banking services.

5.7 VIRTUALIZATION AND CLOUD COMPUTING

Virtualization is the source of technology used for activating cloud computing. It is the process of disintegration of a single physical server into multiple logical servers. As the division of the physical server takes place, each logical server starts behaving like a physical server and has the ability to run an operating system and applications independently. Virtualization services are provided by many famous companies like Microsoft, which use their virtual servers for storage instead of PCs. These virtual servers are fast, cheap and save time [7].

Virtualization has proved very useful for software developers and testers, as it allows the developer to write a platform-independent code and to check it.

It is mainly used for three purposes:

(i) *Network virtualization*: the process of combining the given resources in a network by dividing up the given bandwidth. Each part of the bandwidth is independent of the others and can be given to a particular server or device in real time.

(ii) *Storage virtualization*: the sharing of physical storage from multiple network storage devices to what looks like a central storage device. This is controlled by a central console. It is commonly used in storage area networks (SANs).

(iii) *Server virtualization*: covers server resources, such as processors, RAM, operating systems, from server users. The purpose is to increase resource sharing and reduce the burden and complexity of computation for users.

Types of virtualization are shown in Figure 5.6.

Virtualization is the key to unlocking the cloud system because of its ability to decouple the software from the hardware. For example, virtual memory can be used by PCs to borrow extra memory from the hard disk. The substitution of virtual disks for real memory works perfectly if properly managed, even though virtual disks are slower than real memory. They can also imitate a whole computer, which means a single computer can perform the function of 20 computers.

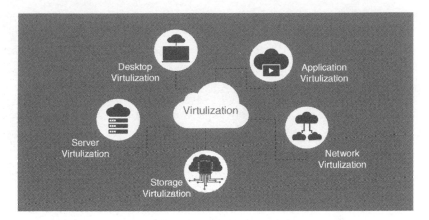

FIGURE 5.6   **Virtualization.**

5.8 TECHNIQUES

Techniques designed to enable grid computing can be applied to the cloud. Their purpose is to share data from different resources with different users.

5.8.1 Service Orientation and Web Services

This covers implementation of the operational aspects and administrative functions. Service orientation offers businesses the opportunity to make changes in the scalability of the cloud. In some ways, the grid faces challenges in building an informatics infrastructure and expanding administrative functions such as security and job submission [9].

5.8.2 Data Execution

Data managed across a cloud platform can be shared across the public and private clouds. Data stored in the cloud has its own data security rules. The benefits of using cloud data management are:

- Connection of processes such as disaster recovery

- Ransomware protection for keeping data secure

- Cost-effectiveness

Some data-intensive applications require easy access to information. In the grid, data management takes place on a large scale. GridFTP is the most common data transfer protocol in grid computing. Single application programs are not applicable to high-level activity in which multiple applications can be served in parallel to all computers [11, 19].

5.8.3 Monitoring

Monitoring tools have already been developed to deal with the high-level information provided in cloud computing. This complex process can be carried out using various systems and resources. Grid Monitoring Architecture, a model for grid computing, was developed by the OGF working group [11, 12]. The performance of data is fixed and useful for grid activity. Monitoring serves the following purposes:

- Provides debugging motive

- Full utilization of resources

- Security

- Management decisions

- Performance checking

Data are updated more frequently than they are requested. This means that tools must be able to respond quickly before the data expire. Among monitoring tools are Ganglia, Network weather service, etc.

5.8.4 Amazon Cloud Watch

It keeps a check on Amazon Web Services cloud resources such as Amazon EC2. It finds data from services and transforms the information into readable data that is stored for two weeks. This makes it available to the user, and overall demand patterns such as CPU utilization, network traffic, etc., are also available [2, 8, 9].

5.8.5 Azure Diagnostic Monitor

This finds data in local memory, diagnoses it and transfers it to an Azure database for permanent storage. Transfer of data takes place on demand [2, 8, 9, 13].

5.8.6 Hyperic Cloud Status

This application detects performance in physical, indirect and cloud infrastructures. It provides open-source monitoring and administration software for all types of applications, for example, Amazon Web Services. It produces real-time reports and infrastructure measurements [2, 13].

5.8.7 Autonomic Computing

Autonomic computing is a self-managing computing model. It can control the functioning of applications and systems without data requests from users, aiming to create complex high-level systems that can run automatically and efficiently. An autonomic system is capable of making decisions to respond to the changes at run time, as shown in Figure 5.7. Some important and valuable features are:

- It collects and aggregates information to support user decisions

- Self-management

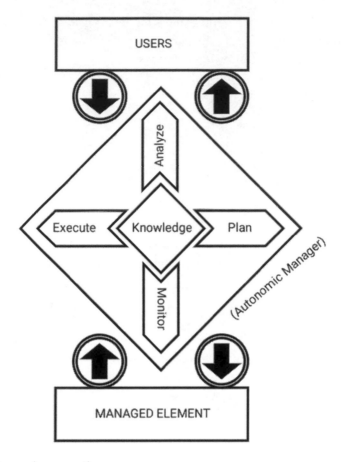

FIGURE 5.7   Autonomic computing.

The four aspects of self-management are:

Self-configuration: it automatically configures the components;

Self-healing: it automatically discovers and corrects faults in the data;

Self-optimization: it automatically monitors and controls data for excellent functioning;

Self-protection: it identifies threats and protects from further attacks [2, 11, 13, 14].

5.8.8 Structure and Effects of Clouds and Grids

Since it is not physically possible or feasible to evaluate and compare each and every stage of grid computing, simulations allow us to research large configurations with high resource demands that can easily be modified. Simulation models include:

- CloudSim: a tool kit that enables the simulation and creation of virtual machines (VMs) on stimulated data centre nodes. It enables the study of associated policies for the transfer of VMs for the reliability and performance of applications.

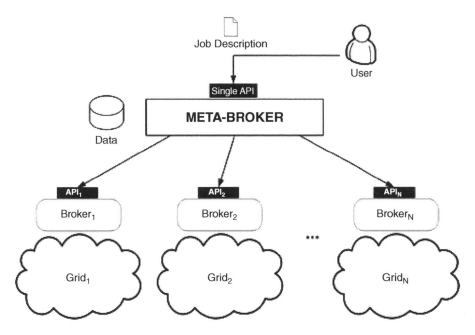

FIGURE 5.8   Interoperability.

- Opnet is a commercial network simulator which manages application performance to satisfy business demands.

Simulations are not only very important for cloud computing research, but also address aspects of the various layers within it.

5.8.9 Interoperability

Interoperability and portability are the abilities to build a system to bring reusable components together and make them work together. A tool that takes this approach into consideration is meta-brokering, which is used by various projects [10, 20]. Mature grid technologies can analyze activities in the field of interoperable clouds, as shown in Figure 5.8.

5.9 THE FUTURE OF GRID COMPUTING

Grid computing methods began in response to demand for high-performance computing, but they also come with some disadvantages. Nevertheless, they are a widely used computing medium requiring minimal changes in the configuration of the CPU. There are two major steps in the parallel implementation of all the methods of progressive alignment:

- Pairwise alignment for construction of tree structure
- Progressive multiple alignments

Parallel T-Coffee, the first parallel implementation of T-Coffee, works in two main stages:

- The first stage lists pairwise alignments that are formed and stored in the database.

- In the second stage, the evaluation of sub-alignment takes place in the database.

The two stages work in parallel depending upon the transfer of messages and access to the remote memory. This property of parallelism, together with the distribution of pairwise alignment, speeds up the process of scheduling during library generation. Parallel versions of statistical alignments are also available. For example, MSAProbs works by (i) calculating pairwise posterior matrices; (ii) constructing a guided tree from those statistics; (iii) performing the transformation.

5.10 CONCLUSION

Grid operations and technology are some of the most appropriate evaluations in today's smart world [17]. Grid is a heterogeneous, loosely coupled and geographically distributed computing system in which servers distribute pieces of data and collect results from clients. A grid combines the processing capabilities of different computing units and uses them together to complete a particular task. However, security can be a major issue with grid computing.

The emerging technology of cloud computing is descended from grid computing. Grids and clouds are similar in many areas like architecture, technology and technique, but cloud enjoys more advantages. Various organizations are opting for cloud computing as it provides multi-purpose benefits that grid computing cannot show. The technologies generated by grid computing have sped up the growth of the cloud, creating brand-new virtualization opportunities.

REFERENCES

[1] Gupta, Er Piyush. *A study of grid computing and cloud computing*, 2014. scribd.com.
[2] R. T. Nakhate and G. D. Korde. The role of grid computing technologies in cloud computing. *International Journal of Research in Advent Technology*, 2(2), February 2014. E-ISSN: 2321–9637.
[3] K. Gurudatt, S. Vikas, G. Shyam, P. Mitra, and T. Pune, *Grid computing overview*, 2013.
[4] P. Srivastava and R. Khan. A review paper on cloud computing. *International Journal of Advanced Research in Computer Science and Software Engineering*, 8:17, 2018. 10.23956/ijarcsse.v8i6.711.
[5] C. Restrepo, J. Pérez, J. Aranda, and J. Diaz. Towards formal interaction-based models of grid computing infrastructures. *Electronic Proceedings in Theoretical Computer Science*, 144, 2014. 10.4204/EPTCS.144.5.
[6] M. Ayala-Rincon, E. Bonelli, and I. Mackie. *Proceedings 9th International Workshop on Developments in Computational Models. Electronic Proceedings in Theoretical Computer Science*, 2014. 144. 10.4204/EPTCS.144.
[7] K. Vieira, A. Schulter, C. Westphall, and C. Westphall. Intrusion detection for grid and cloud computing. *IT Professional*, 12(4):38–43, July–Aug. 2010.

[8] R. Buyya, C. S. Yeo, S. Venugopal, J. Broberg, and I. Brandic. Cloud computing and emerging IT platforms: vision, hype, and reality for delivering computing as the 5th utility. *Future Generation Computer Systems*, 25:599–616, June 2009.

[9] S. Sotiriadis, N. Bessis, and N. Antonpoulos. Decentralized meta-brokers for inter-cloud: modeling brokering coordinators for interoperable resource management. In *2012 9th International Conference on Fuzzy Systems and Knowledge Discovery*, Sichuan, 2462–2468, 2012.

[10] S. Sotiriadis, N. Bessis, and N. Antonpoulos. Decentralized meta-brokers for inter-cloud: Modeling brokering coordinators for interoperable resource management. In *Proceedings—2012 9th International Conference on Fuzzy Systems and Knowledge Discovery, FSKD 2012*, 2012. 10.1109/FSKD.2012.6234313.

[11] S. Y. Junfeng and W. Chengpeng. Cloud computing and its key techniques, 320–324, 2011. 10.1109/EMEIT.2011.6022935.

[12] R. Sharma, V. K. Soni, M. K. Mishra, and P. Bhuyan. A survey of job scheduling and resource management in grid computing. *World Academy of Science, Engineering and Technology International Journal of Computer and Information Engineering*, 4(4), 2010.

[13] C. Quan and D. Qianni. Cloud computing and its key techniques. *Journal of Computer Applications*. 29.25622567.2009.

[14] Z. Zhao, C. Gao, and F. Duan. A survey on autonomic computing research. In *2009 Asia-Pacific Conference on Computational Intelligence and Industrial Applications (PACIIA)*, Wuhan, 288–291, 2009.

[15] Essays, UK. *Application of cloud computing in various sectors information technology essay*, November 2013. Retrieved from https://www.uniassignment.com/essay-samples/information-technology/application-of-cloud-computing-in-various-sectors-information-technology-essay.php?vref=1.

[16] S. Zhang, X. Chen, S. Zhang, and X. Huo. The comparison between cloud computing and grid computing. In *2010 International Conference on Computer Application and System Modeling (ICCASM 2010)*, Taiyuan, V11-72–V11-75, 2010.

[17] J. Dawn Thompson. Future perspectives: high-performance computing. *Statistics for Bioinformatics*, 97–107, 2016. doi:10.1016/b978-1-78548-216-8.50010-5.

[18] N. Mustafee. Exploiting grid computing, desktop grids and cloud computing for e-science: future directions. *Transforming Government: People, Process and Policy*, 4(4):288–298, 2010. https://doi.org/10.1108/17506161011081291.

[19] M. Elhoseny, A. S. Salama, A. Abdelaziz, and A. Riad. Intelligent systems based on cloud computing for healthcare services: a survey. *International Journal of Computational Intelligence Studies*, 6(2/3):157–188, 2017

[20] M. M. Wang, Z. G. Qu, and M. Elhoseny. Quantum secret sharing in noisy environment. In X. Sun, H. C. Chao, X. You, and E. Bertino, Eds. *Cloud Computing and Security. ICCCS 2017*, Springer, Cham. Lecture Notes in Computer Science, vol 10603, 2017. https://doi.org/10.1007/978-3-319-68542-7_9.

CHAPTER 6

Application of Networks in Cloud Computing

Karuna Middha, Anjali Chaudhary, and Prayag Tiwari

CONTENTS

6.1   INTRODUCTION TO CLOUD NETWORKING

The pay-per-use model has proved very effective in helping government agencies, enterprises, educational institutions and others to modify the on-demand capacity of the IT infrastructure and to minimize its operational and maintenance costs by utilizing resources and applications from the cloud. An enterprise can have its own virtual infrastructures in the cloud, enabling its users to utilize the concept of "resource pooling" – also referred to as *private cloud*. These private clouds are customized according to user requirements and supported by an environment that is firewall protected and is different from the cloud that provides general services to the public – known as *public cloud*. These public cloud services provide users with Infrastructure as a Service (Iaas), Platform as a Service (PaaS) and Software as a Service (SaaS). Examples include Amazon EC2 and Google App engine. Private and public clouds combine to generate a *hybrid cloud*, and a a huge number of enterprises prefer to have their infrastructure created on hybrid clouds. The architecture of

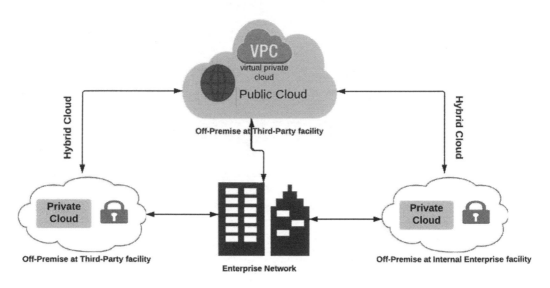

Fig. Private, Public & Hybrid Clouds

FIGURE 6.1 Public, private and hybrid clouds.

a hybrid cloud can be considered as a system in which a private cloud expands its coverage area to merge with a third party (e.g., a public cloud environment) to make use of additional resources securely and according to demand [1, 2] (Figure 6.1).

Since networks are a crucial component of an IT infrastructure, and a cloud can be regarded as an online IT infrastructure, networking is a key concept in cloud computing. This chapter introduces the different types of networks, and the operating systems that provide and manage network services and devices. Operating system functions and how these support cloud computing will be presented through a discussion of network architecture. The importance of networks in cloud computing and existing challenges in cloud networking will be explained. The chapter concludes with a discussion of the future scope of networking.

6.2 NETWORKS

A network is a link between devices/computers that share data with each other. All the devices (printers, computers, storage devices) that are connected to a network are known as *hosts*, and each device has a *network interface card* (NIC) that is connected to another device via a network cable or another communication medium. The communication medium/network cable transfers information in the form of electrical pulses to and from the hosts.

When multiple hosts are connected via a single network, a device called a *switch* is used to further distribute information to other connected hosts. *Routers* are used to connect multiple networks together.

6.2.1 Types of Network

There are different types of network (Figure 6.2):

- *Local Area Network (LAN)*: A network that is created within a room, a floor or a building such as in the home, school or college laboratory.

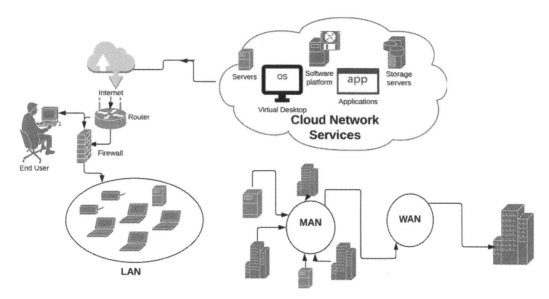

A LAN, MAN, WAN & Cloud-Based Network connecting to an individual user and a LAN network

FIGURE 6.2   Types of networks.

- **Wide Area Network (WAN):** A network that is highly scalable, spans a larger area than LAN and consists of many networking devices connected together, such as on a university campus.

- **Internet:** A system that is formed by connecting many LANs and WANs together. Internet service providers (ISPs) connect the internet to LANs. Point of presence (POP) is the access point through which ISPs communicate with a regional network.

- **Cloud-Based Network:** An enterprise or an organization can distribute its network around the globe using a cloud-based network created by cloud service providers. For its part, the enterprise has to connect its own network to the cloud network for global connections. In the cloud-based network, the enterprise can provide a multi-user application to serve multiple users while their data remain separate and invisible to other users. Maintenance and operational tasks are very simple in the cloud network. Thousands of employees can work together on different modules of a single application without interfering in each other's tasks, integrating the system to yield the final product.

6.2.2 Network Operating Systems

An operating system (OS) is system software that manages all other software applications and hardware on a computing device. When we talk about operating systems in cloud computing, we refer to the OS that is installed on a cloud platform. The choice of operating system decides the overall functionality of a cloud-based service and how efficiently it provides infrastructure to its users. Most operating systems work on a computer's hard drive, but the cloud operating system works on a remote server. With cloud services, the client just needs to install the application on their device as an interface to the required service, and all the data is saved and processed only by the remote server [3].

There is a wide variety of cloud operating systems:

Chrome OS: A client-based operating system, working like a thin web client, providing access to data stored on the cloud and other online applications. Since internet speed is the limitation, this operating system is currently limited to lightweight web applications, but with advancements in technology and broadband speeds, it will increasingly be working with heavier applications as well (Figure 6.3).

Glide OS: Offering 30GB free storage capacity, Glide OS could be an even better choice for team projects since it allows six members within an account. Glide OS provides its users with automatic file and application compatibility across devices and operating systems. The Glide OS also provides the Glide Sync App, which helps to synchronize home and work files (Figure 6.4) [1].

ZeroPc: ZeroPc, developed by Zero Desktop Inc. (California), is a cloud-based operating system that is easily accessible through a web browser, with a mobile application that can be easily installed on smartphones/tablets running on Android, iOS and Amazon's version of Android. This OS is quite efficient in bringing together a large collection of photos/media from a huge number of cloud-based platforms to a single place, as it can easily transfer files from one cloud platform to another storage platform. There is also an inbuilt media player in this OS [4] (Figure 6.5).

FIGURE 6.3 Chrome OS.

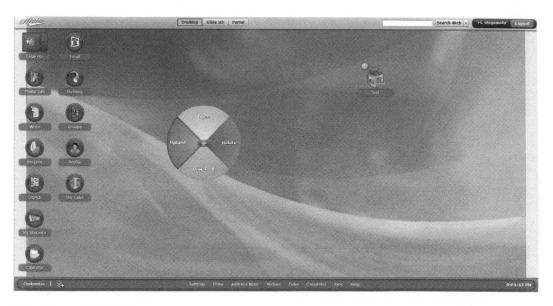

FIGURE 6.4   Glide OS.

FIGURE 6.5   ZeroPc OS.

The Place A OS: A cloud-based operating system with the capability to add web applications like radio, file manager, calendar, notes, etc. as links that can be opened in new tabs. This OS also provides instant messaging features to its users. It somewhat resembles the Linux OS which also has a quick response rate (Figure 6.6).

FIGURE 6.6   The Place A OS.

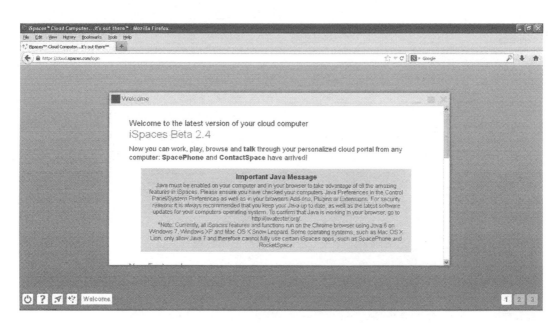

FIGURE 6.7   iSpaces OS.

iSpaces: A multi-desktop cloud computer that runs on standard browsers and supports all major operating systems. Working environments can be effectively organized by getting easy access from one workspace to another (Figure 6.7).

6.2.3 Network Architecture

This section briefly discusses network architecture, design and implementation, the networking process and the major components in a network system.

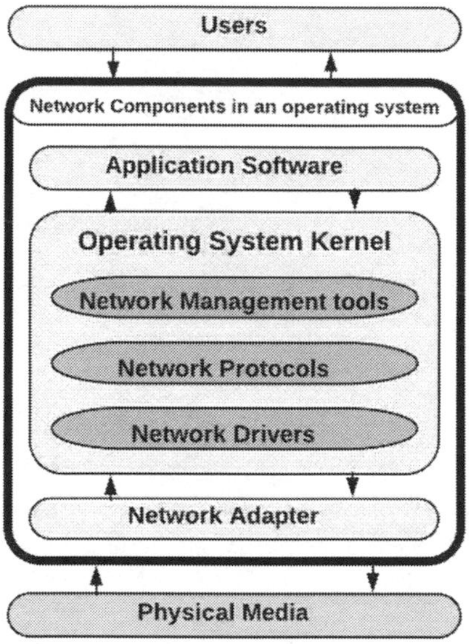

FIGURE 6.8   Network components.

A network can range from two computers connected via copper wire to millions of computers and peripherals connected through the internet. A network between devices is formed using major components in layers (Figure 6.8).

The application software acts as a service interface between user and operating system, which then takes charge of data transmission to the physical channel, which further connects to other devices in the network. Within the application software, the user's requests for database access, file transfers and message passing are processed and submitted to the operating system, where data requested is collected by network management tools to decide which protocol is to be used for data communication.

A network protocol is a pre-defined set of rules that determines the process of data transmission and acts as a service interface between network drivers and application software. A network protocol takes charge of tasks such as error detection, data formatting, establishing communication ports, deciding the destination route by locating the desired destination, managing network traffic, the data transmission and receiving process, authentication, data encryption, and so on. A network driver is an interface between the hardware and the software of a computer system, enabling the operating system to establish communication with the NIC to connect with physical data transmission media. It also handles I/O interrupts during data transmission and interacts with protocols, network adapters and buffers.

To connect the physical media to another computer system in the network, we use a network adapter, which can either be a wired ethernet NIC or a wireless device that communicates data in the form of binary electric signals and frames to specify transmission rate and signal strength. It works on the handshaking principle, instructing the operating system to receive further information once the previous information is received by a peer adapter [1].

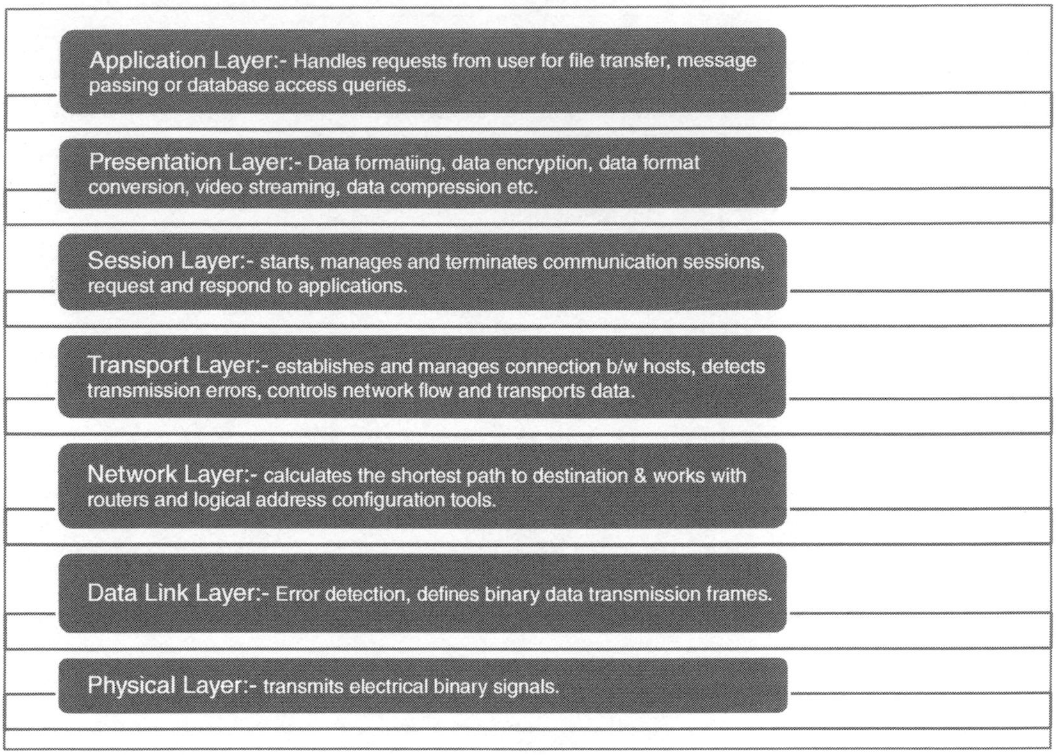

Application Layer:- Handles requests from user for file transfer, message passing or database access queries.

Presentation Layer:- Data formatiing, data encryption, data format conversion, video streaming, data compression etc.

Session Layer:- starts, manages and terminates communication sessions, request and respond to applications.

Transport Layer:- establishes and manages connection b/w hosts, detects transmission errors, controls network flow and transports data.

Network Layer:- calculates the shortest path to destination & works with routers and logical address configuration tools.

Data Link Layer:- Error detection, defines binary data transmission frames.

Physical Layer:- transmits electrical binary signals.

FIGURE 6.9   OSI network architecture.

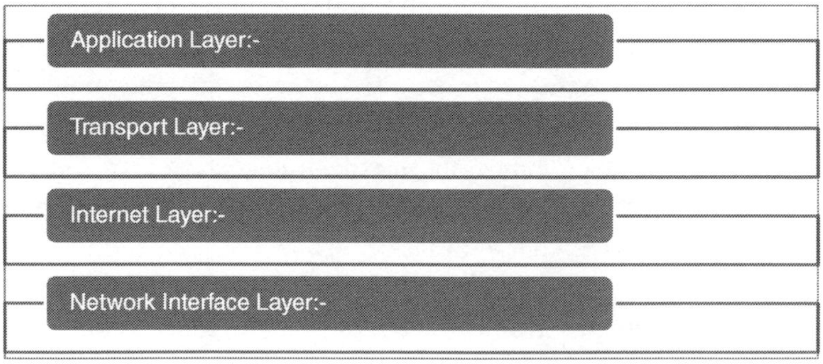

Application Layer:-

Transport Layer:-

Internet Layer:-

Network Interface Layer:-

FIGURE 6.10   Internet architecture.

There are two different architectures to represent a network.

- *OSI* (*Open System Interconnection*) architecture developed by the International Organization for Standardization (ISO) to define the process of communication between two computers through seven layers (Figure 6.9).

- *TCP/IP* (*Transmission Control Protocol/Internet Protocol*), also known as the internet architecture, consists of four layers (Figure 6.10) and is designed to implement the

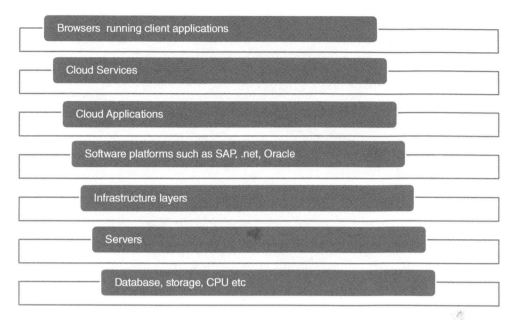

FIGURE 6.11   Cloud architecture layers.

exchange of data through the internet. The internet architecture contains the application, presentation and session layers of the OSI model. The transport layer of both architectures is equivalent, and the internet layer of the TCP/IP model is similar to the network layer in the OSI model. The network interface layer in the internet architecture includes the data link and physical layer of OSI architecture.

6.2.3.1 Cloud Architectural Layers

There are multiple layers in the architectural implementation of cloud computing. The topmost layer represents the browsers that run on desktop and mobile platforms to access the multiple applications being hosted in the virtual cloud environment (Figure 6.11). The next layer constitutes the services and applications used by cloud users according to the pay-per-use model of cloud computing.

These services and applications require some software platforms to run, which form the fourth layer of the architecture of the cloud (Oracle, .net, SAP, etc.). In what is one of the most amazing innovations of technology, the bottom layer of this cloud computing architecture is the infrastructure that manages storage, databases, servers and CPU.

6.2.3.2 Network Architecture for Clouds

Cloud architecture "encompasses a variety of systems and technologies, as well as service and deployment models and business models" [5].

It is a combination of three sub-architectures – business, technical and operational – which are defined as follows:

- Business architecture is used to design business aspects and covers various cost models, service contracts and pricing. It is also responsible for setting up appropriate communication with the stakeholders of the business model.

- Technical architecture works on the design of various cloud components. It considers all the security and structural aspects of the cloud platform to make the service available to the maximum number of users.

- Operational architecture monitors all the operational tasks being performed while providing cloud services and establishing cloud connections in a virtual environment.

There are three fields of the cloud-based computing network architecture which it is important to explore:

- The *data centre network* enables the connection of infrastructure elements such as servers and storage within a data centre.

- The *data centre interconnect network* architecture connects various data centres.

- The *internet services* connect users to cloud service providers.

Internet services concerns architecture and infrastructure of telecommunication services currently available in the market, a vast field. Hence, we confine ourselves in this chapter to exploring the other two fields of network architecture.

6.2.3.2.1 Data Centre Network

A data centre network (DCN), which uses a hierarchical network design during its construction, is known to provide highly scalable and reliable cloud services to its end users through huge data centres that incorporate hundreds and sometimes thousands of servers and other storage devices (Figure 6.12).

The bottom layer of the DCN, i.e., the access layer, is responsible for ensuring connectivity between server resources within the data centre. Factors such as server virtualization, form and density of servers play a vital role in the design of this access layer. There are various design approaches to the connectivity of access layers, depending upon server hardware and the application requirements. They include ToR (top-of-rack) switch, integrated switches (blade server chassis consisting of blade switches arranged in a modular fashion or in the form of an embedded software switch) and EoR (end-of-row) switch.

The middle layer of the DCN, i.e. the aggregation layer, is a strengthening point between the connected switches of the access layer that provides server connectivity for multi-tier applications along with core network connectivity for the internet, clients or the WAN. Generally, this aggregation layer is the borderline between routed links (layer3) and ethernet broadcast domains (layer2) within the data centre. 802.1Q VLAN trunks are used to connect the aggregation layer to the access layer of the DCN, since it allows the same physical switch to be connected to multiple servers with different IP subnets and VLANs.

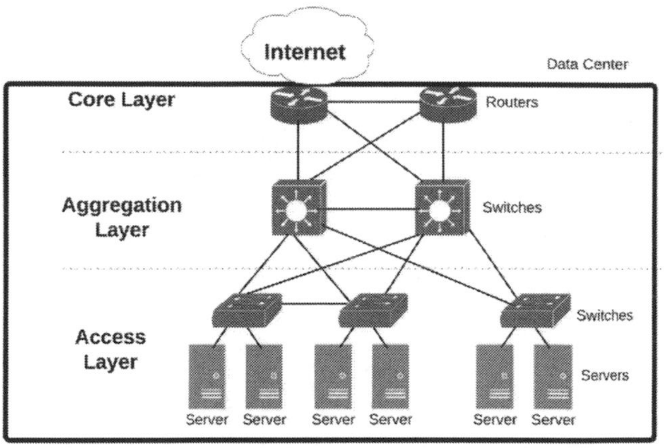

FIGURE 6.12   Data centre network architecture.

The top layer of the DCN, i.e., the core layer, provides efficient layer3 switching to manage IP traffic between the telecom service provider's internet edge and the data centre. In situations where the cloud service provider has their data centres connected through a metropolitan area network (MAN) or a privately owned WAN network, it would be a better choice of network architecture design to expand the layer2 networks for multiple data centres. In other situations, the public internet carries the traffic, and in such cases it is preferable for the core switches of the data centre to be connected via peering routing topology (layer3). This configuration of connecting links as layer3 point-to-point connections provides rapid convergence in case of link failures and prevents the exposure of the control plane of switches through end node devices to broadcast traffic network.

Evidence shows that networking technologies are evolving mostly at the access layer of the DCN because of the increasing number of servers within a data centre. According to recent research, the concept of switch virtualization is a common approach allowing the logical access layer (layer2) to span its functionalities across various physical devices. Various architectures exist to implement the concept of switch virtualization – including virtual ethernet switch, fabric extender and virtual blade switch (VBS) technologies – enabling flexible configurations and multi-tenant cloud computing services. These are highly scalable according to client requirements and workloads .

In addition, there are some cloud service providers who are implementing two-tier architecture (front-end and back-end tiers) in data centres for the optimization of cost and delivery services.

6.2.3.2.2 Data Centre Interconnect Network
Data centre interconnect networks (DCINs) provide connectivity between different data centres to ensure a better and more efficient cloud computing experience for their clients (Figure 6.13). Although the conventional concept of connecting various data centres in a virtual private network (VPN) offered a secure means of communication and high reliability, DCIN has turned out to be a more efficient networking class in cloud computing

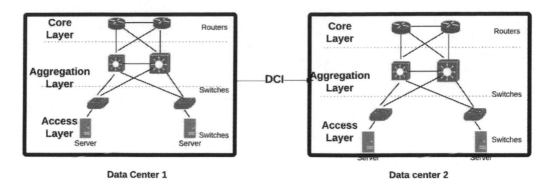

Data Center 1   **Data center 2**

FIGURE 6.13   Data centre interconnect.

services. It works by extending layer2 networks in multiple data centres. For example, it would be more convenient to maintain the adjacency of layer2 networks in servers across various data centres rather than renumbering the IP addresses of the servers in case of server migration (either due to pre-planned data centre maintenance or unplanned dynamic workload balancing).

Extending the layer 2 network is cost-effective and fulfils business needs, since it allows maintenance of operational policies and similar server configuration. The most important technical requirements of DCIN include application mobility for multiple sites, dynamic workload balancing, data centre disaster avoidance, high-availability clusters and virtual server migration(dynamic). With the evolution of technology in the field of cloud computing and networking, continuous improvements and research are needed in order to connect DCNs more efficiently [6, 7].

6.2.3.2.3 Network Protocols
Protocols are a set of rules that allows devices to communicate with each other in a network. The following are some of the different types of protocols that are used in cloud computing.

- *Gossip Protocol*: A communication protocol, also referred to as an epidemic protocol, in which the anti-entropy technique used for repairing data is distorted. A dissemination protocol is used to spread information, using a flooding agent in a network. This protocol is used for failure detection, for messaging, using a TCP to wrap data, and for monitoring proper transmission.

- *Connectionless Network Protocol* (*CLNP*): A datagram network protocol that works on the third-layer protocol of the OSI model. Its fundamental working principle is similar to the transport layer in internet protocol. It provides sufficient mechanisms for fragmentation (data fragment/total length, data unit identification, information verification by computing checksum on the header of CLNP and offset).

- *State Routing Protocol* (*SRP*): Consists of routers that exchange messages so that every router can learn the network topology based on which each router computes its routing table using the shortest-path computation, for example, IP, internetwork packet exchange (IPX).

- *Internet Group Management Protocol* (*IGMP*): A communication protocol that provides a mechanism for a computer to multicast its content to other nodes in the network via routers. It uses the OSI model and is a part of the network layer.

- *Secure Shell Protocol* (*SSP*): A cryptographic network protocol to provide more reliable network services such as remote command execution and command-line login. This offers proper authentication to the users in a network.

- *Coverage Enhanced Ethernet Protocol* (*CEE*): This protocol is used to resolve issues of packet traffic within the data-link layer, since it offers lower-cost data storage within packets.

- *Extensible Messaging and Presence Protocol* (*XMPP*): This protocol is based on XML (extensible mark-up language), which allows structured data to be exchanged in the nearly real-time environment. This protocol is also designed for files, video transfer and IoT applications, such as smart grid, gaming and much more.

- *Advanced Message Queuing Protocol* (*AMQP*): An application layer protocol whose features are message orientation, information packet queuing, maintaining security and reliability. It is a wire-level protocol, which describes the data format in the form of a byte stream to be sent across the network.

- *Enhanced Interior Gateway Routing Protocol* (*EIGRP*): An advanced routing protocol used to automatically make and configure routing decisions. It only sends incremental updates that reduce the workload of routers.

- *Media Transfer Protocol* (*MTP*): An extension to the picture transfer protocol (PTP) through which media files can be transferred to and from portable devices over a cloud. It cannot be used to transfer video files.

6.2.4 Network Security

Along with high-tech services and leisure possibilities brought to us by cloud technologies, there are also threats, and our assets, whether internal or in the cloud, become vulnerable to attack since this network segment has internet access (Figure 6.14).

Among the threats to a network connected via the internet are session hijacking, malicious domain name service (DNS) responses, XSS (cross-site scripting), MITM (man in the middle), allowing botnet commands and botnet zombies.

It is therefore a priority for network designers and service providers to establish a system through which network elements can support outbound interactions via the internet and inbound interactions to be handled by the controlled sharing segment (DMZ). Effective use of cryptography allows a secure network to safeguard users' assets and private information. Firewalled subnets are used to provide protection for both inbound and outbound interactions, while network address translation (NAT) is used to provide data abstraction

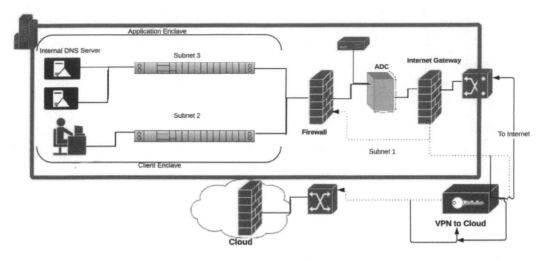

FIGURE 6.14   Secure internet access application example.

capabilities. VPNs provide the foundation for secure and authenticated communication between user resources and cloud resources. In an application delivery controller (ADC), multiple mechanisms for network protection are integrated into a single network device that can prevent data loss and data leakage by providing protection at application levels. Other security mechanisms incorporated by service providers include the hardware security module (HSM) to provide cryptographic acceleration, DNS, and the security information and event management (SIEM) system which allows alerts and audits to support input coordination of security systems and intrusion detection and prevention systems (IDPs) along with the ADC.

Security features of cloud infrastructure include identity and access management, data encryption, segregation between the users and protection, VM isolation, secure VM migration, virtual network isolation, security intelligence and software, platform and infrastructure security, and security event and access monitoring. Security ensures:

- Authentication and authorization access

- Uninterrupted availability

- Maintenance of client confidentiality

- Subscriber identity management

6.3 IMPORTANCE OF NETWORKING IN CLOUD COMPUTING

It is the underlying network infrastructure of cloud computing that is the foundation for the super-efficient services and applications being provided to its users. Although quite complex in structure, networking is the key principle enabling the internet to bring intelligent and intellectual features to its customers. Network intelligence and connectivity allow multiple services like data, videos and music to be accessed via a single platform where security and management services are all merged.

With the rising trend towards establishing a more secure end-to-end automated service with the least manual interference, networks need to be made more reliable and efficient to meet the demands of users. For the same reason, our networks are becoming programmable, ensuring flexibility to grow and shrink with changing consumer demand. Some of the network-based innovations being discussed include overlay transport virtualization (OTV), virtual port channels (vPCs), fabric path and locator/ID separation protocol (LISP), among many others. Therefore, networks play a crucial role in cloud computing and are periodically modified, raising our living standards and business profits beyond what we might expect.

6.4 CHALLENGES IN EXISTING CLOUD NETWORKS

To meet the varied demands of cloud services, existing networking architectures work on the one-size-fits-all model. Although topologies, protocols and services are designed with all the requirements in mind, it does not mean maximum usage of cloud services. The following are some of the challenges faced in cloud networking:

- **Less control:** There are many government agencies and industries that feel quite uncomfortable with the idea that their data will be located somewhere on a system that is not under their control. Hence it becomes a challenge to the service providers to design a network that ensures security transparency to its customers [8].

- **Application performance:** There should be a guaranteed bandwidth allocated to every application that is hosted on the cloud network to avoid delays and latency on the client side and ensure timely performance to the users.

- **Flexible deployment of appliances**: Security appliances like intrusion detection systems (IDSs), deep packet inspection (DPI) and firewalls are deployed by many enterprises to protect their applications from malicious attacks. It therefore becomes crucial to enable applications hosted on the cloud to function flexibly without violating the functioning of such appliances [9].

- **Security management:** Cloud service providers must be able to supply virtual control systems to the client to manage security and firewall settings for runtime environments and applications in the cloud system.

- **Policy enforcement complexities:** It becomes extremely necessary to enforce policies like traffic isolation and controlling access to clients, which requires topologies, protocols and configuration of routers and switches to be altered. This appears to present a tough challenge to network designers.

- **Location dependency:** Location dependency constraint arises because servers and appliances are tied to a static physical network. This dependency constraint limits the flexibility and resource utilization of a virtual cloud network.

6.5 FUTURE SCOPE IN CLOUD NETWORKING

Cloud service providers are operating their own data centres at present. But emerging technologies and user demand may lead to situations where an enterprise, an organization or an end user asks for the capability to work with multiple cloud providers simultaneously because of local access migration between service providers or due to competitive services being given to them. This could give rise to increased demand for cloud interoperability, also known as cloud federation. Cloud federation manages consistency and access controls when two or more independent cloud computing facilities share either authentication, computing resources, command and control, or access to storage resources [10].

6.6 CONCLUSION

This chapter has covered the basic aspects of networks in cloud computing – network types, operating systems, protocols, architecture, challenges and the future scope of networks – illustrating the importance of networking as the basis for the services and applications provided by the cloud environment. The never-ending opportunities and innovations in the field of cloud computing will definitely lead us in unanticipated directions as we seek to make optimum use of the intelligence and resources available to mankind.

REFERENCES

[1] Chao, L. (2016). *Cloud Computing Networking*. New York: Auerbach Publications. https://doi.org/10.1201/b18867.

[2] Bharti, D. and Goudar, R. (2012). Cloud Computing–Research Issues, Challenges, Architecture, Platforms and Applications: A Survey. *International Journal of Future Computer and Communication*. 10.7763/IJFCC.2012.V1.95.

[3] Bardhan, N. et al. (2015). *International Journal of Computer Science and Information Technologies (IJCSIT)*, 6 (1), 542–544. http://ijcsit.com/docs/Volume%206/vol6issue01/ijcsit20150601121.pdf.

[4] Desktop, Z. (n.d.) *Zeropc - Your content navigator for the cloud*. [online] Zeropc.com. Available at: http://www.zeropc.com/.

[5] NIST Special Publication 800-146. (May 2012). *DRAFT Cloud Computing Synopsis and Recommendations, Recommendations of the National Institute of Standards and Technology*. https://nvlpubs.nist.gov/nistpubs/Legacy/SP/nistspecialpublication800-146.pdf.

[6] Furht, B. and Escalante, A. (2010). *Handbook of Cloud Computing*, 1st ed. Boston, MA: Springer Science+Business Media, LLC. DCN.

[7] Systems, C. (2009). *Cisco data center interconnect design and implementation guide*. System Release 1.0. [online] USA: Cisco Systems, Inc. Available at: https://www.cisco.com/c/dam/en/us/solutions/collateral/data-center-virtualization/data-center-interconnect/data_center_interconnect_design_guide.pdf.

[8] IBM Research – Zurich, Christian Cachin (April 2011). https://cachin.com/cc/talks/metis2011.pdf.

[9] Azodolmolky, Siamak et al. (2013). Cloud computing networking: Challenges and opportunities for innovations. *IEEE Communications Magazine*, 51, 54–62.

[10] Sridhar, T. (September 2009). Cloud computing - A primer, part 2: Infrastructure and implementation topics. *The Internet Protocol Journal*, 12(4), 2–17. http://www.cisco.com/web/about/ac123/ac147/archived_issues/ipj_12-4/124_cloud2.html.

CHAPTER 7

Privacy and Security Issues in Cloud Computing

Akanksha Kochhar and Anubha Khanna

CONTENTS

7.1 INTRODUCTION

7.1.1 Definition

Cloud computing is the on-demand service of resources that are used through the internet. These services include all the software, servers, networking, storage, databases, and so on. It does not require service providers and is manageable with very little effort [1].

Cloud computing is a new technology that provides flexibility at very low cost. While still an evolving technology, it is becoming very popular and rapidly gaining market share [1]. To make cloud computing successful, it is crucial to develop efficient and optimal solutions.

Typically cloud computing consists of the following five elements:

1. *On-demand self-service.* The user can automatically and instantaneously access the resources without needing to interact with providers.

2. *Broad network access.* Resources are easily deliverable over the network, and customers can use them on different platforms, such as laptops, smartphones or PDAs.

3. *Resource pooling.* Various resources are combined for customers using different models, for example, virtualization or multi-tenancy [2]. Resources can be dynamically assigned or reassigned according to customer needs. In this model some physical resources are completely invisible to the user, who has no control over location or formation.

4. *Rapid elasticity.* All the resources are easily and immediately available to customers. No contract is required.

5. *Measured service.* Metres can be used to measure the available resources, even though these are used and grouped by multiple users.

7.1.2 Service Models
Cloud computing works on three service models, which are explained below.

7.1.2.1 Infrastructure as a Service (IaaS)
IaaS is a demand-based service that lets the user pay for what they use, reducing costs.

Advantages:

(a) **Continuous and disaster recovery**: IaaS has access to various data centres and applications in various locations.

(b) **Scaling up or down**: Resources can be scaled up or down based on the demand in any cloud computing group.

(c) **IT focus**: IaaS focuses on IT infrastructure more than on any business activity. It is best for an entrepreneur that can manage and own their infrastructure. Any testing can then be done on that owned infrastructure.

7.1.2.2 Platform as a Service (PaaS)
PaaS builds a platform for customers to expand and manage applications without the need to consider the associated infrastructure. It will work in cooperation with IaaS.

Advantages:

(a) It greatly reduces complexity by allowing a high level of programming.

(b) Applications can be built effectively by providing various hardware and software tools.

(c) PaaS gives more emphasis to software development.

Authentication can be appropriately checked when data is being transferred, which ensures the integrity of any application.

7.1.2.3 Software as a Service (SaaS)

This is a platform in which licensing must be provided for software and models used. Licences are granted on a subscription basis and the system is fully centralized.

Advantages:

(a) **Customization**: Any customer can change the settings to modify the look and feel of the application.

(b) **Rapid delivery**: Applications built by SaaS will deliver very quickly – in months or even weeks – due to central hosting, as the developer will update everything without involving the customer at all.

(c) **Cooperative functionality**: SaaS will work with users in collaboration to share data and information.

In open-source SaaS, the source code is freely available to the customer. It is a web-based application where the service provider provides all services.

7.1.3 Deployment Models

Networking, software infrastructure, platform and storage are services that can be increased or decreased according to market demand.

Deployment models can be classified as follows:

Private Cloud. This term has recently been introduced to describe cloud computing in private networks. An internal data centre is built within the organization [3]. Scalable resources and different virtual applications are pooled by cloud vendors and made available to cloud users. Higher-level privacy and security can be achieved by the private cloud through various firewalls and operations performed on networks.

Public Cloud. A public cloud is one that can be available to every user who wants to access the public network. This network makes resources accessible to the public through the internet. Security is a significant issue in public clouds.

Hybrid Cloud. This model aims at providing all the services available in both public and private clouds. It can incorporate links in one model by the user and in the other by the third party. A hybrid cloud provides greater security than public clouds. It is a complete package forming an automated, flexible, cost-effective and well-managed platform for network models.

Community cloud. This cloud model can be implemented by many organizations at the same time but with a similar concern in terms of security requirements, goals and policy considerations. This type of cloud uses one or more organizations and is managed and controlled by a third party [4, 5].

7.2 RESEARCH CHALLENGES IN CLOUD COMPUTING

Cloud computing consists of applications, platforms and infrastructure. Each aspect performs its task and produces the desired results for business and personal tasks. Business applications consist of SaaS, PaaS, IaaS, web services, service providers, utility computing and service commerce [6].

As cloud computing deals with integrated technologies, many systems are involved, including networking, database management, operating systems, virtualization, resource scheduling, transactional control, concurrency control, load balancing and memory management. Security issues apply to all these systems, and security systems need to be maintained in a cloud computing environment.

Many security challenges have been addressed and solutions have been implemented, including data encryption. However, a number of challenges remain for research in this field:

- **Service-level agreement (SLA)**

  The cloud needs to be administered by an SLA, which covers issues like data protection, cost and outages. Customers also expect the SLA to provide back-up and data archiving and protection. The SLA allows many instances of an application to be copied onto multiple machines, according to a priority system. A major challenge for cloud customers is to calculate the SLAs of cloud vendors.

- **Data management**

  The amount of data in the cloud is enormous; it may be in structured or unstructured form. Service providers have to rely on infrastructure providers for data security as they do not have access to the physical security system [8]. The infrastructure provider has to achieve confidentiality and auditability for security. Confidentiality of data means the data must be accessed and transferred securely by applying cryptographic protocols. Auditability involves the use of remote attestation protocols to ensure security applications are not tampered with.

- **Data encryption**

  The main technology for ensuring data security is encryption. Before uploading a file to the cloud, data needs to be encrypted to protect it from unauthorized access.

- **Migration of virtual machines**

  Many programs run on a machine using virtualization, or one program is run on more than one machine. Virtualization can be used in cloud computing to balance load in the data centre. The advantage of virtual machine migration is to avoid hotspots [9].

- **Interoperability**

  Interoperability means the ability to work for more than two systems so that information can be exchanged [10]. Some public cloud networks are designed in such a way that they are not able to interact with one another. This design fault prevents organizations from combining their IT systems in the cloud to improve cost and efficiency. To overcome this challenge, the standards of the organizations must be improved so that cloud service providers can design interoperable platforms to enable data portability.

- **Bandwidth cost**

  High-speed communication channels are important for the efficient operation of cloud computing. Efficient hardware and software can be developed to save on costs, but bandwidth is expensive [11]. The front input cost can still be reduced by migrating data to the cloud, but the network cost is increased. A major issue arises if data is consumers' private data and if it is distributed to various clouds. Data-intensive jobs must therefore only be undertaken on private clouds [12].

- **Virtualization**

  Virtualization is the creation of a virtual version of the storage devices, operating system, multiple servers or networks, enabling the customers to migrate their data to a remote destination. It divides tasks between multiple environments, and data abstraction of resources is carried out to simplify the system. There are two types: base metal virtualization (Type 1 hypervisors) and operating system virtualization (Type 2 hypervisors). Virtualization provides elasticity, scalability, independence of location, cost-effectiveness and a simple interface. But the challenges are workload, security, unnecessary migration, automation and on-demand elasticity.

- **Energy consumption**

  Cloud data centres which set up the infrastructure contain many servers. These servers consume a lot of energy, are very expensive to operate and generate excess heat which needs to be removed [7]. The ultimate goal is not only to reduce energy consumption but also to maintain environmental standards.

- **Management and scheduling of resources**

  Management of resources can be done at various levels – hardware, software and virtualization – to verify parameters like performance, security, management of memory space, CPU, threads and device controllers. Job scheduling is an aspect of resource provisioning which aims to optimize the turnaround time, arrival time, burst time, waiting time, throughput, response ratio and utilization of resources. Cloud computing comprises many technologies to which job scheduling strategies can be applied.

  Scheduling of jobs is an important process. Incorrect scheduling may have serious consequences and lead to wastage of resources.

- **Reliability and availability**

  The strength of technology can be measured in terms of the degree of its reliability and availability. Reliability means how often the resources are available without disruption. To achieve a reliable system, resource utilization must be performed

[13]. Availability means that resources can be obtained at any time when needed. However, even a reliable and available system may be subject to service denial, slow performance or natural disasters.

- **Scalability and elasticity**

  The two main features of cloud computing are scalability and elasticity. The users of these features make unlimited use of these services according to their needs. Scalability is the ability of the system to perform well, even if the resources are scaled up or down. Elasticity is the ability to scale the resources up or down when required. Elasticity allows the static integration and extraction of the resources in the infrastructure [14]. Scalability can be horizontal or vertical: horizontal scalability means adding more nodes in the form of new machines to existing systems, or the addition of memory or processors to an existing computer.

- **Accessibility of servers and applications**

  Traditionally administrative access to servers was controlled and restricted to direct or on-premise connections. But in cloud computing, administrative access is through the internet, which increases risk. Administrative access is important for maintaining changes in system control. Accessing data is mainly related to the security concerns which the user faces in accessing the data. Security policies may need consideration where individuals are not given access to certain data [15]. To avoid data intrusion by unauthorized users, these security policies need to be followed.

- **Data integrity**

  Data integrity means that the contents of the data should not be tampered with. Corruption of data can occur at any time during storage and at all levels. Transaction control protocols such as ACID (atomicity, consistency, isolation, durability) are used to ensure the integrity of data.

- **Data availability**

  If the cloud depends on a single service provider, it may suffer from data unavailability in the case of system failure. A multi-tier architecture involving several servers is therefore required. Even if one system is unable to provide the available data, others are still available to service the cloud.

- **Data segregation**

  Data present in the cloud is in a shared environment. Encryption of data may not be the only solution, as encryption techniques may destroy the data. If encryption is done, it should be carried out at different levels by experienced personnel [16].

- **Protection of data storage**

  Protection of data is one of the significant security issues in cloud computing. The best and most secure technique is cryptographic encryption followed by self-shipping of the key of the encrypted hard drive.

- **Patch management**

  The self-servicing nature of cloud computing may make patch management difficult. If an enterprise subscribes to a resource, the patch management for that enterprise will become the responsibility of the subscriber. Cloud computing systems should therefore maintain the patch resources supplied by the vendor.

- **Security policy and compliance**

  Service providers are required to adhere to audits (such as ISO standards) and security certification. If these standards are not maintained, customer trust in security will be adversely affected. The organization implementing the audit must check and continue to monitor contractual, legal and policy requirements.

7.3 CLOUD COMPUTING SOLUTIONS

7.3.1 Security Issues

The cloud environment faces major security challenges. Research is ongoing in this area.

7.3.1.1 Trusted Third Party

The security solution to preserve the integrity, authenticity and confidentiality of data in a cloud environment is trusted third party (TTP). TTP facilitates secure interactions among the parties and those parties' trust in this third party. Developing trust between the two parties or establishing an assurance of trust will result in acceptance [19]. TTP provides end-to-end services to the information systems sector and is used in different specialist areas of industry. The process operates in a top-down manner, so every layer needs to build up trust with a secure adjacent layer [17]. Security is required at the operational, technical, procedural and legal levels. TTP certificates are regarded as a reliable passport and can establish a secure connection with the user.

A TTP can guarantee trust if the parties follow the certification rules and confirm their credentials. An end user can do this via a digital certificate which authenticate themselves with the cloud service. The digital certificate together with the service provider's certificate can be used to create a secure SSL connection [18]. The application provider can use the certificate to authenticate the credentials and to perform encryption and decryption of data. Servers and devices also use a digital certificate to communicate securely.

The third party reviews the communications between the other two parties to prevent fraudulent activities. This party provides business confidence in electronic form, enabling both parties to securely transact data. Digital certificates are distributed in such a way that they can be associated with the servers at the original and the destination sites that are participating in the transfer of information. The following processes can be implemented in a technically satisfactory and logically acceptable way via the public key infrastructure (PKI):

Authentication: This process involves the identification of the parts which are included in an electronic transaction and transfer of information by digital means. Figure 7.1 depicts the authentication process in a cloud environment.

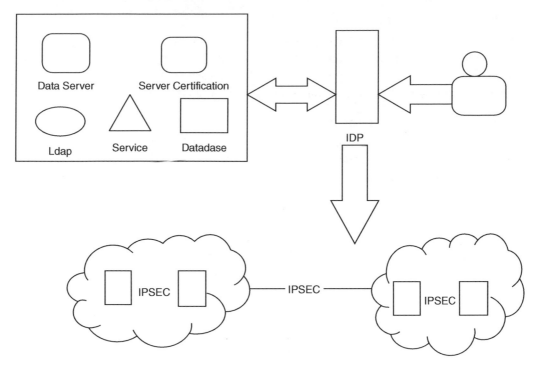

FIGURE 7.1 Authentication process in the cloud environment.

Authorization: Access to resources, databases and information systems must be authenticated.

Confidentiality of data: The data must remain protected from unauthorized access when transferred to a different location or stored in a database.

Integrity of data: The data must remain protected from unauthorized modification when commuted to a different location or stored in a database.

Non-Repudiation: When the transaction is completed, no party can deny having sent the data.

PKI, which is a distributed information system, has the advantage of coupling with the directory. A directory is defined as the set of objects with attributes that are similar and organized logically or hierarchically. The Lightweight Directory Access Protocol (LDAP) is the standard way to access directory services, which are limited to the X500 data model.

When a directory is joined with PKI it can distribute:

- Certificates, where an end-user certificate is obtained before sending the encrypted message.

- Certificate status information.

- Private keys, when portability is needed in an environment where users will not use the same server every day. The directory will store the encrypted keys, which then will be decrypted by the user using the password.

PKI can be deployed with single sign-on (SSO) mechanisms, which are good for a distributed environment like a cloud environment within the boundaries of which the user can navigate. The user has to sign in using a password or may use any authentication mechanism to access the resources on different servers.

TTP relies on the following features.

7.3.1.1.1 Low- and High-Level Confidentiality

Transferring data through a communications network in a secure way is a complex issue which brings threats of data interruption or modification. This complexity is further increased in the cloud environment where there is no physical connection.

To maintain secure connections, the PKI implements a secure socket layer or IPSec. IPSec is an IP layer protocol which helps to send and receive packets that are cryptographically secured without any modification. The services provided by IPSec are confidentiality, authenticity and integrity [20]. IPSec users can authenticate themselves using PKI certificates. SSL protocol generates end-to-end encryption, and helps the client–server authentication interface between applications and the TCPIP protocols.

Protection is needed for communication not only between users and host but also between host and host. Whether SSL or IPSec is required depends on the individual needs. IPSec is compatible with any application, but an IPSec client needs to be installed on every remote device. In contrast, SSL is installed on every web browser, so there is no further requirement for client software. Multiple platforms are supported by cloud computing, so users do not need to install the IPSec client for encrypting data. SSL is considered advantageous for user-to-host communication; IPSec is regarded as the most efficient option as it involves host-to-host communications.

7.3.1.1.2 Authentication of Server and Client

Authentication certificates are mandatory in the cloud environment to certify the entities involved in the interactions. Certification includes physical infrastructures, servers, network devices and users (Figure 7.2). The authority responsible for these certificates is PKI certification. To authenticate the user automatically, the use of private keys is recommended, which maintains transparency of the connection between servers and devices. All the services used in the cloud require a strong authentication and authorization to maintain security and digital signatures are recommended in the authentication process to generate the flexibility and mobility for cloud users. Single sign-on (SSO) is a process whereby users who need to use applications deployed on the virtual machines do not need to repeat the authorization process every time they use each service [21].

A standard piece of open-source software called Shibboleth is used to provide SSO. This software, which makes secure authorization decisions for access to resources, depends on

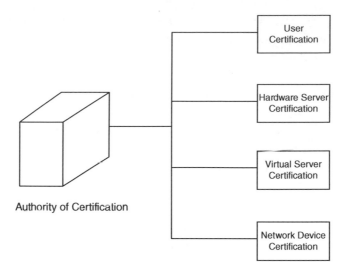

FIGURE 7.2 Types of certificates.

the third party providing data about the specific user. It is very important to differentiate between the authentication and the authorization procedures. In the authentication procedure, the user is requested to navigate to the organization and authenticate himself. In this phase only useful data is transferred. After authentication of the user, access to the resource in the cloud is either granted or rejected, or the process is aborted. During this process, no personal information about the user is threatened. The use of standards like Security Assertion Markup Language (SAML) helps to maximize interoperability among the parties which are transferring the information. This language is an XML-based standard used for exchanging authentication and authorization of data with security issues in mind.

7.3.1.1.3 Formation of Security Domains

With PKI and LDAP, trust relationships can be formed within the parties. A group of all the parties applying the same set of policies and rules to access the resources in the cloud is called a federation. It provides a structure and a framework that follows the authentication and authorization rules. The federations consist of small units of clouds that interoperate and exchange data and resources within defined boundaries. All the clouds present in the federation are independent and can exchange resources with any cloud but in a standardized framework.

7.3.1.1.4 Data Cryptographic

Cryptographic separation of data, processes and computations needs to be hidden from the outside world [22]. Encryption helps to provide confidentiality, integrity and privacy. Symmetric and asymmetric cryptography offers efficiency and security.

7.3.1.1.5 Authorization Based on Certificate

The cloud environment is assumed to be a virtual network of different individual areas, with an ad-hoc and flexible relationship between users and resources. Providers and users of the resources do not have the same security area. So the old identity-based models are no longer efficient or effective. Authentication certificates that need to be issued by a PKI can be used in a web-controlled environment [23]. Access controls, which are attribute based, make decisions based on the attributes of the environment and resources, making the distributed system flexible and scalable.

Two methods can handle the control of access and security concerns related to the cloud.

7.3.1.1.5.1 Access Control Method

Role-based access control (RBAC) is one of the powerful general approaches to managing security. The cloud-based RBAC algorithm comprises four key components: cloud user, access permission, role and session.

- **Cloud user**: the client side of the cloud.

- **Access permission**: the mode granting the right to manipulate data in the cloud.

- **Role**: the collection of permissions required to operate functions in the cloud.

- **Session**: the type of relation which associates one user of the cloud with several roles.

When the session starts, cloud users can request acquisition of roles. If the request is granted, the role will be defined. In this way, data can be protected from malicious attack by access being denied. Cloud-based RBAC defines the algorithms that will correlate with one another.

Cloud user role assignment (CURA) maps sessions to cloud users. Role permission assignment (RPA) maps one session with several roles. Both these algorithms use the cloud-based RBAC component and provide secure access control. If the disabled state is chosen by the cloud user, no role will be activated. This is how data are locked until the cloud user wishes to make use of it. RBAC is flexible and simple to use to control access.

RBAC can be extended – known as credential-based RBAC – to provide the necessary constructs to capture requirements. There is also location-based RBAC (LBAC) and generalized temporal RBAC (GTRBAC). In cloud computing, ISPs have no knowledge of users before access, so the user role cannot be directly assigned in the access control policy. The policy provided by RBAC increases the capability of the cloud computing environment.

7.3.1.1.5.2 Policy Integration Method

If multi-policy problems occur, a flexible control method is used, which eliminates strict boundaries between different policies. Different cloud service providers (CSPs) have to conform to this policy first.

A lot of work has been done on multi-domain policies and issues reported during the integration of policies in the interests of providing a comprehensive framework to ensure efficiency. Researchers have adopted different mechanisms for building security in the cloud, including a centralized strategy focused on providing access and authentication that would be appropriate for all cloud service users. If requirement gathering is dynamic, then a centralized approach may need to become decentralized, as users might need to interact with the environment for the same purpose. The solution for secure interoperation could be SAML, web services standards or Extensible Access Control Markup Language (XACML). Role-mining mechanisms can also be used to define the roles of the existing configuration system. They work by first granting permissions to existing systems and then combining them into a role. Users may take roles based on the different domains according to the service they need. If existing roles were changed, they might disrupt the organization and inhibit the functionality of the system; role mining plays a vital role by providing closed or optimal solutions for existing system roles in a cloud environment. The users in this method will be given two secure policies, policy X and policy Y. A cooperative policy evaluation (CPE) will evaluate the two policies, following which policies will be integrated.

If the integrated policy is not optimal, then a generalized back-up policy Z will be implemented, and it will be regarded as the temporal policy for the ordering of data. To improve trust levels, feedback to the CPE must be allowed, in which the CPE can flexibility adjust the criteria to optimize the evaluation.

7.3.2 Privacy Issues

Research on privacy of data in the cloud identifies two solutions to the issue of its unauthorized use:

7.3.2.1 Identity Management

To prevent the unauthorized use of data, a user-centric approach to identity control has been developed [24]. The cloud privacy label (CPL) is added to identity management to protect cloud users. Here every user has been allotted several attributes for authentication of their identity, which are managed by the user themselves. Privacy protocols are used to check the attributes. In cloud computing, the CSP and the cloud users set a CPL which only allows access to data from the cloud to authorized individuals.

Identity management (IDM) is used to retain the various unique identities. The identifiers can be used to identify and define the user. This will reduce the use of IDM from industry as the user will able to control their own identity digitally [25, 28]. So the user has complete control over their functions. Cloud computing users can acquire data from different locations. Thus there is a need to securely export and transfer the data over the network. IDM ensures that the semantics of the content of the user's information should be maintained. Research into IDM solutions for cloud computing continues.

7.3.2.2 User Control Methods

Cloud data is virtualized so cloud users face certain issues of data loss. A private cloud is one solution, but it can only support small-scale users. So public or hybrid clouds and third-party audit (TPA) between the CSP and the cloud is needed. Capability is granted through TPA. Its role is also to check whether the CSP is maintaining the data storage. In this way, users can control and manage their important data. In this method, cloud users produce verification data to give to TPA, which will be sent for auditing. Then polling of the states of data is done by TPA, and finally, the data is analyzed. In this way, the users have control of the data and can develop more trust with the CSPs [26].

7.4 CONCLUSION

Security and privacy are among the major concerns in the cloud computing environment. This chapter defined cloud computing and discussed service and deployment models. Security and privacy challenges faced by the cloud were also discussed, together with some solutions.

REFERENCES

[1] T. Dillon, C. Wu, and E. Chang. Cloud computing: issues and challenges. In *2010 24th IEEE International Conference on Advanced Information Networking and Applications*, Perth, Western Australia, 2010.

[2] P. Mell and T. Grance. *Draft NIST working definition of cloud computing – v15*, August 21, 2009.

[3] M. Armbrust, A. Fox, R. Griffith, A. Joseph, R. Katz, A. Konwinski, G. Lee, D. Patterson, A. Rabkin, and I. Stoica. Above the clouds: a 32 Berkeley view of cloud computing. *Tech. Rep. UCB/EECS-2009-28*, EECS Department, University of California, Berkeley, CA, 2009.

[4] L. Badger, T. Grance, R. P. Comer, and J. Voas. *DRAFT cloud computing synopsis and recommendations—Recommendations of National Institute of Standards and Technology (NIST)*. National Institute of Standards and Technology, Gaithersburg, MD, May 2012.

[5] IBM Global Services. *Cloud computing: defined and demystified explore public, private, and hybrid cloud approaches to help accelerate innovative business solutions*, April 2009.

[6] R. P. Padhy, M. R. Patra, and S. C. Satapathy. Cloud computing: security issues and research challenges. *IRACST – International Journal of Computer Science and Information Technology & Security (IJCSITS)*, 1(2), December 2011., 136–138.

[7] V. Krishna Reddy, B. Thirumal Rao, L. S. S. Reddy, and P. Sai Kiran. Research issues in cloud computing. *Global Journal of Computer Science and Technology*, 11(11), July 2011.

[8] X. Zhang, N. Wuwong, H. Li, and X. J. Zhang. Information security risk management framework for the cloud computing environments. In *Proceedings of 10th IEEE International Conference on Computer and Information Technology*, Bradford, 2010, pp. 1328–1334.

[9] C. Wang, Q. Wang, K. Ren, and W. Lou. Ensuring data storage security in cloud computing. In *17th International Workshop on Quality of Service*, Charleston, SC, July 13–15, 2009, pp. 1–9. ISBN: 978-1-4244-3875-4.

[10] European CIO Cloud Survey. *Addressing security, risk, and transition*, May 2011.

[11] A. Leinwand. *The hidden cost of the cloud: bandwidth charges*, July 2009.

[12] J. Gray. Distributed computing economics. *ACM Queue*, 6:63–68, May 2008.

[13] H. Wu, Y. Ding, C. Winer, and L. Yao. Network security for virtual machines in cloud computing. In *5th International Conference on Computer Sciences and Convergence Information Technology*, Seoul, November 30–December 2, 2010, pp. 18–21. ISBN: 978-1-4844-8567-3.

[14] G. T. Lepakshi. Achieving availability, elasticity, and reliability of the data access in cloud computing. *International Journal of Advanced Engineering Sciences and Technologies*, 5(2):150–155, April 2011.

[15] K. Hwang, S. Kulkarni, and Y. Hu. Cloud security with virtualized defense and reputation-based trust management. In *Proceedings of 2009 Eighth IEEE International Conference on Dependable, Autonomic and Secure Computing (Security in Cloud Computing)*, Chengdu, China, December 2009, pp. 621–628. ISBN: 78-0-7695-3929-4.

[16] M. D. Dikaiakos, D. Katsaros, P. Mehra, G. Pallis, and A. Vakali. Cloud computing: distributed internet computing for IT and scientific research. *IEEE Internet Computing Journal*, 13(5):10–13, September 2009. DOI: 10.1109/MIC.2009.103.

[17] D. Polemi. Trusted third-party services for health care in Europe. *Future Generation Computer Systems*, 14:51–59, 1998.

[18] S. Castell. *Code of practice and management guidelines for trusted third party services*. INFOSEC Project Report S2101/02, 1993.

[19] Commission of the European Community. *Green paper on the security of information systems*, ver. 4.2.1, 1994.

[20] A. Alshamsi and T. Saito. *A technical comparison of IPSec and SSL. Cryptology, 2004. Proceedings of the 19th International Conference on Advanced Information Networking and Applications (AINA'05)*, 1550-445X/05, © 2005 IEEE.

[21] Cloud Identity Summit. *Secure the cloud now, Cloud identity summit*. Retrieved on November 10, 2010 from http://www.cloudidentitysummit.com/.

[22] C. P. Pfleeger and S. L. Pfleeger. *Security in Computing*. Prentice-Hall, 2002.

[23] B. Lang, I. Foster, F. Siebenlist, R. Ananthakrishnan, and T. Freeman. Attribute-based access control for grid computing, 2008. *J Grid Computing* (2009) 7:169–180. DOI 10.1007/s10723-008-9112-1.

[24] M. Ko, G.-j. Ahn, and M. Shehab. Privacy enhanced User-Centric Identity Management. In *IEEE International Conference on Communications*, Dresden, Germany, 2009.

[25] C. Wang, Q. Wang, K. Ren, and W. Luo. *Privacy-preserving public auditing for data storage security in Cloud Computing*. IEEE Communication Society, 2010.

[26] Z. Wang. Security, and privacy issues within the Cloud Computing. In *2011 International Conference on Computational and Information Sciences*, Chengdu, China.

[27] M. Elhoseny, A. Abdelaziz, A. Salama, A. M. Riad, A. K. Sangaiah, and K. Muhammad. A hybrid model of Internet of Things and cloud computing to manage Big Data in health services applications. *Future Generation Computer Systems*, 86:1383–1394, September 2018. DOI: 10.1016/j.future.2018.03.005.

[28] M. M. Wang, Z. G. Qu, and M. Elhoseny. Quantum secret sharing in noisy environment. In *Cloud Computing and Security. ICCCS 2017*. Lecture Notes in Computer Science, X. Sun, H. C. Chao, X. You, and E. Bertino, Eds., Vol. 10603. Springer, Cham, 2017. DOI: 10.1007/978-3-319-68542-7_9.

CHAPTER 8

Design and Development of E-Commerce–Oriented PaaS Application in Cloud Computing Environment

Samridhi Seth, Rahul Johari, and Kalpana Gupta

CONTENTS

8.1 INTRODUCTION

Cloud computing allows people to access data and applications over the internet by removing the barriers of system-specific storage. It has the advantage that people can access data from anywhere, anytime, on any device with the help of the internet. To provide dynamism in storage, a large number of devices are connected over a network. All companies are moving towards cloud computing, since it reduces investment costs, increases profit, is easy to maintain, carries almost negligible maintenance costs, requires less hardware and storage spaces on site, and provides backup facilities. Further, most importantly, it is both secure and location independent, therefore it has much reduced overheads.

In cloud computing, data, applications, services and software are stored and accessed over the internet on a secure channel, minimizing the possibilities of data loss. Using the cloud, we can create new applications and services, back up data, host websites and blogs, and analyze data. In everyday life, it is used for storage of data, for example, in Google Cloud. For the sake of convenience, we can explore the various benefits of using cloud computing as described in references [1–6].

Popular open-source cloud computing tools are detailed in Table 8.1. Additional benefits of using cloud computing are:

- One pays for the amount of data accessed/used.

- Clouds provide self-service access, that is, an account is created.

- Explores the features of virtualization to the fullest.

- Easy sharing of resources.

- Easy migration from one cloud platform to another.

- Things can easily become viral.

- Reduced need to focus on infrastructure and space security.

- It helps in work automation.

- Accessible on different platforms such as mobile, laptops, etc.

- Resource pooling.

- Clouds systems are organized in such a way that they automatically optimize resource usage, providing transparency to systems.

- User preferences based on the usage can be saved.

- Accessible at different locations.

TABLE 8.1 Open-Source Cloud Computing Tools [7, 8]

S. No.	Tool Name	Year Launched	Tool Description
1.	Eucalyptus	2008	– Private and hybrid cloud-driven deployments
2.	OpenNepula	2008	– A service management tool
			– Manages heterogeneously distributed infrastructure
			– Data centre virtualization platform
3.	Open Stack	2010	– PaaS management tool
4.	Cloud Stack	2013	– Offers public and private cloud-based services
			– IaaS-based component.
			– Supports majority of the cloud computing features
5.	Synnefo	2015	– Provides computer, network, imaging, volume and storage services
			– Manages Ganeti clusters at back end
6.	FOSSCloud	2011	– Helps build private or public cloud
			– Supports virtualization
7.	OpenQRM	2004	– Provides automated workflow engine
			– IaaS component
8.	CloudFoundry	2011	– Container-based application development
			– Fast and efficient
9.	Salt stack	2011	– Configuration management software
			– Orchestration tool
10.	Core OS Fleet	2014	– Container based
			– Configuration management software
11.	Google's Kubernetes	2015	– Automated deployment, scaling and management
			– Container based
			– Orchestration tool
12.	Apache Mesos	2016	– Cluster manager
			– Resource isolation and sharing
13.	Smart OS and Smart Data Centre	2011	– Use analytics and IOT
			– Lightweight virtualization tool

(*continued*)

TABLE 8.1 (Continued) Open-Source Cloud Computing Tools [7, 8]

S. No.	Tool Name	Year Launched	Tool Description
14.	Ansible	2012	– Used for automation
			– Configuration management
			– Application deployment
15.	WSO2 Stratos	2013	– PaaS component
			– IaaS component
			– Orchestration tool
16.	Cloudify	2012	– Uses a declarative approach
			– Orchestration tool
			– Used for automation
			– Configuration management
			– Application deployment
17.	Tsuru	2012	– Extensible
			– Fast and efficient
			– Orchestration tool
			– IaaS component
			– Recovers failed units automatically
18.	EMUSIM	2010	– Offers simulation and emulation combination
			– Cost efficient
			– CPU-intensive applications supported
19.	CloudAnalyst	2010	– Extended version of CloudSim with some extra features
			– Powerful simulation framework
			– Easy-to-use GUI
			– Efficient output
			– Flexible
20.	iCanCloud	2010	– Simulation of storage networks
			– Easy and flexible
			– Customized virtual machines
			– Easy-to-use GUI
21.	DCSim	2012	– Extensible data centre simulator
			– Event-driven simulator

The cloud computing architecture consists of four layers [12]:

- **Physical layer:** contains servers for multiple uses such as personal and public access.

- **Virtualization layer:** contains virtual machines to connect servers with each other.

- **Networking layer:** used to provide networks and routing.

- **Application layer:** Responsible for services used between end users.

Big data is today gaining greatly increased importance by virtue of cloud computing and cloud storage, leading to the term big-data cloud computing [9]. Big-data cloud computing has brought revolutionary changes to the workings of industry. Every company is moving towards cloud storage and new big-data features such as the Internet of Things and artificial intelligence. The new technology can be easily integrated with current technology. Big-data cloud computing platforms include Google's Cloud Platform, Microsoft Azure, Rackspace or Qubole. It has greatly helped entrepreneurs who can now start their business with just a modest investment. It saves on infrastructure, money and time. If the outcome is not successful, a person can start again without having experienced significant losses.

Crypto cloud computing addresses the security and privacy concerns that arise with cloud computing. Data is stored in encrypted form and each user manages their personal account where access is granted only to the authorized user. Cryptography secures the data with encryption techniques and prevents delay in information exchange as it reduces the load and prevents hackers from attacking. In crypto cloud computing, data is allocated on demand. Crypto cloud computing helps gain the trust of users [13, 14].

A load balancer [16, 17] manages the traffic on the servers, ensuring equal distribution of the workload. Various techniques are used to decide which server the workload is allocated to. Factors used for determining the techniques are: number of requests, number of servers, load on servers, network, server health and predefined conditions. One drawback of load balancing is its overheads. The load balancing algorithm varies, depending upon the position of the load balancer, whether on the network layer or the application layer.

Application-layer algorithms [18, 19] are:

- **Dynamic Round Robin:** Weights are assigned on the basis of real-time calculations.

- **Least Connections:** New services are assigned to those servers having the fewest active connections. Servers are continuously monitored for capacity computation.

- **Weighted Least Connections:** Services are assigned on the basis of the number of active connections and the relative capacity of the server. If two servers have same number of connections, then the one with higher weight is assigned the request.

- **Source IP Hash:** Servers are selected on the basis of their IP address. Each user is assigned a particular server. IP addresses are stored on the servers. A hash key is generated using a unique combination of source and destination IP addresses. Services are only redistributed in the event of a server failure. This scheme is generally used on online shopping sites to retain items in the cart.

- **URLHash:** Hashing is performed on a URL address, removing redundancy and increasing the capacity of the servers as no redundant cache is created at the backend.

- **Least Bandwidth Method:** The server consuming the least bandwidth for the last 14 seconds is selected.

- **Least Packet Method:** The server transmitting the fewest packets is selected.

- **Software Defined Networking Adaptive:** Combines information from different networking layers to decide request allocations. This helps in information sharing about the network and communications.

Applications of cloud computing [5, 6, 10–12] are:

- **Education:** Virtual classrooms and smart classes are enabled via cloud computing. It has changed the way teaching used to take place and has motivated many young children to learn.

- **Industries:** Industry has seen growth since businesses have become transparent and improved their quality of service. Smart technology saves time and has made them more profitable.

- **Healthcare:** Treatment and patient history can all be shared with the team via the cloud.

- **Banking:** Banks have moved to paperless working, with everything online and digitized.

- **Email services:** Nowadays, everything can be stored on Google accounts.

- **Voice calls and video calls:** These smartphone features are accessible via cloud computing.

- **Data storage:** Documents, presentations, etc. can be stored in the cloud and accessed from anywhere, for example, Google documents, spreadsheets.

- **Business:** Many online businesses are flourishing, for example, online marketing.

- **Big-data analytics:** Big data includes voluminous, variable, volatile and vulnerable data which needs to be archived securely. This is possible in the cloud. Many big-data analytics tools are designed for cloud computing.

- **Social networking:** Most social networking sites are based on cloud computing.

- **Free software:** Software available online free of charge is stored in the cloud.

- **Navigation:** Easy to locate addresses and find routes.

- **Disaster recovery:** Easier to recover data from the cloud than from fixed-asset storage.

8.2 LITERATURE SURVEY

Cloud computing is the sharing of resources on the internet. Since all users can access the services once they have created an account, multiple users need simultaneous access to the data, which implies the use of scheduling algorithms. Various meta-heuristic solutions for cloud task scheduling have been proposed. LBACO, a task-scheduling algorithm [20], aims to balance the load on the system. Simulations have been conducted on the CloudSim toolkit package. Other authors have discussed the basis of cloud computing, and reviewed scheduling algorithms that are widely used in cloud computing [21, 22, 24]. A comparative analysis of these algorithms is also undertaken. In [23], the author(s) discuss the challenge of task scheduling, that is, optimized solutions for scheduling tasks in cloud computing. A service-level agreement must be in place and should be renewed on a regular basis.

In [25], the author(s) present a detailed survey of various security aspects of cloud computing. A three-tier security architecture for effective and efficient SaaS- and PaaS-based services is also proposed. Reference [26] describes a scheme using social group organization algorithms for resource allocation and SJF-based scheduling algorithms for task scheduling. Experimental results show improved performances. In [27, 28] uses and applications of cloud computing are reviewed, together with optimization techniques for scheduling algorithms. The authors detail the factors on which computing is based, including service-level agreements, CPU utilization, costs, and so on.

Security and data privacy have been major concerns in the implementation of cloud computing techniques, with everything available on the internet and cyber-crime reaching new levels with digitization. In [29], the author(s) survey security threats and areas of concern related to security, including user access management, recovery support and data protection. Cloud computing architecture is also discussed. Service-level agreements between two parties are described in [30]. These agreements provide more security as they also lists rules and limit the accessibility of the data between the two parties by describing the guarantees offered to subscribers. Further challenges in this field of research are also discussed.

Another important concept in cloud computing is the deployment of applications on clouds. Issues encountered during deployment cover networks, storage and infrastructure. In [31], cloud computing architecture and different deployment models are explained and resource allocation strategies are listed with their advantages and disadvantages. A proposed open-source deployment system called AURA [32] helps to track failure issues and proposes recovery solutions. AURA is based on Python and works on graph algorithms.

In [33], author(s) analyze the costs of a multilevel cloud computing security model. Cost and performance of different methods of transfer delay are investigated. The algorithm used is Markovian-process algebra PEPA. The results show that predictions can be made on steady states for a given system with some limitations.

Cloud computing and big data are mutually necessary In [34], security, efficiency and privacy are discussed in relation to big data and the Internet of Things. An architecture based on firewall security is proposed, using a security wall between the cloud computing server and the internet server. In the proposed system, interaction between the objects is via wireless media.

In [35], an LB-PSOGSA algorithm is proposed which focuses on load balancing of virtual machines. Virtual machine processing speed and length of the work are the factors affecting load. A policy premium payment application was designed and developed in a Java programming language, showcasing safe financial transaction on the web [36]. The application was successfully deployed in the cloud. Google App Engine, a Google cloud, was used effectively for the design and deployment of the application. In [37, 38] a new "Cyclic Cryptographic Technique" and "Unique identification number (UID)" application is presented which has been skillfully designed so as to fully exploit the power of cloud computing.

In [39] the nuances of BCD are explained: big data, cloud computing and distributed computing. In [40] definitions of cloud computing, cloud services, application deployment over the cloud, and platforms and infrastructure available for cloud computing are attractively presented. In [41] there is a discussion of the widespread great benefits of the extremely cheap and very safe and secure cloud-based environment now available across all IT entities. Cloud computing has changed traditional ways of working in factories and industry. The sharing of limited resources and availability on demand of high-speed computational power has made the job easy for both small and medium enterprises.

8.3 PROPOSED FRAMEWORK

The objective of the proposed research work is to design and develop a state-of-the-art distributed computing PaaS-based application deployed in the cloud computing environment.

Mobile Recharge, a PaaS payment application, has been programmed in Java with the sole purpose of carrying out secure financial/commercial transactions over the web. The application is designed and deployed using Google App Engine, a Google cloud. The proposed system uses encryption techniques to encode the original information.

8.3.1 Steps involved in Verification and Validation of Application

The proposed framework consists of two steps:

1. A Java-based PaaS financial transaction OLTP application is designed. The application receives the following inputs before data is transmitted for verification and validation on the secure payment gateway:

    a. Credit card number

    b. CVV [card verification value] number

    c. Expiry date.

2. The application is programmed in Java using Eclipse IDE (see Figures 8.1–8.4).

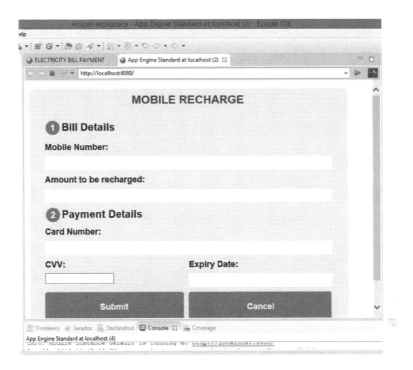

FIGURE 8.1   Application developed on Eclipse.

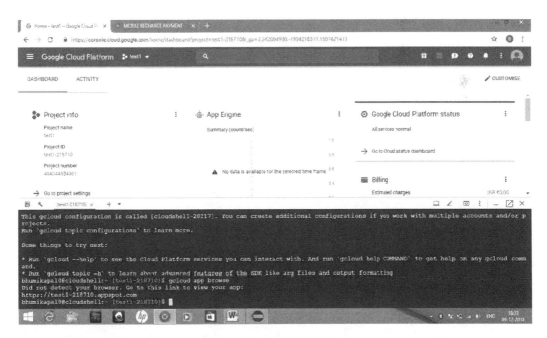

FIGURE 8.2   Application deployed on Google Cloud Platform.

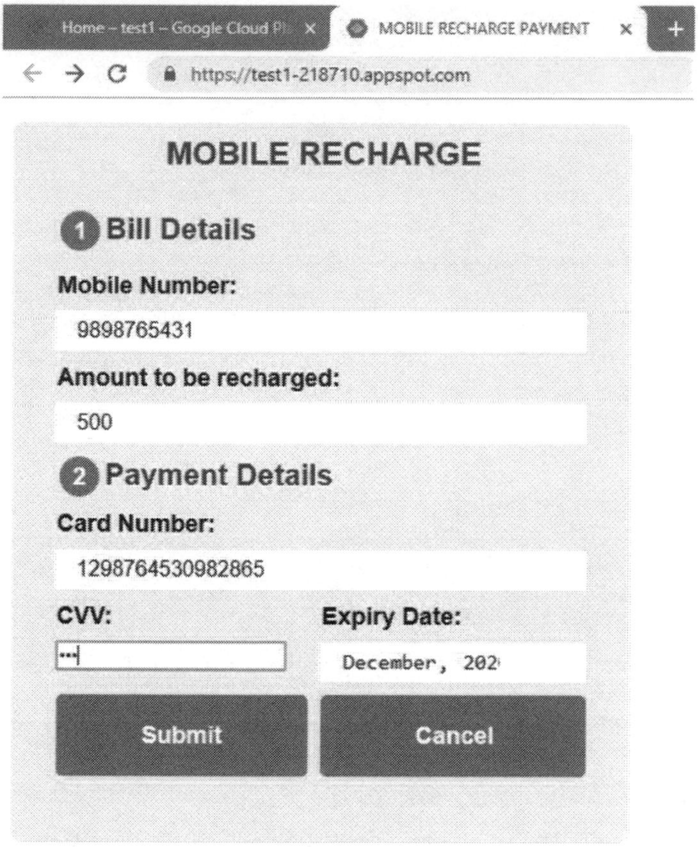

FIGURE 8.3   Application running on server.

FIGURE 8.4   Result shows mobile has been recharged.

8.3.2 Web Application

Web applications are those that can be accessed via a web browser over a network as shown in Figure 8.1. Web applications are needed for business-to-business (B2B) interactions. They are deployed at the server side on the network and executed at the client machine.

Many online payment methods are used by customers, such as debit card, credit card, internet banking and mobile payment. Online financial transaction processing between two parties in what is known as e-commerce takes place through a payment gateway, a process of interface with the web-based business website and the merchant account. Security, encryption and other critical inputs are the most important factors in this process.

8.3.3 Credit Card Validation Process

- **Credit card:** A payment card that may be used by cardholder to borrow money or buy services or products on credit.

- **Credit card number:**

  1. The first digit of the credit card number represents the major industry identifier:

     - 1 and 2 – airlines

     - 3 – travel and entertainment

     - 4 – VISA

     - 5 – MasterCard

     - 6 – banking

     Together with the next five digits of the credit card number, the first six digits form the Issuer Identification Number.

  2. The next nine digits (7 to 15) of the credit card number represent the unique customer account number.

  3. The last digit is the check digit, used to validate the credit card number.

- **CVV number:**

  The card verification value is a 3-digit number.

  Expiry date: The date after which the card will not work. Credit cards have an expiry date because their magnetic strips would not work forever [42].

8.3.4 LUHN MOD 10 Algorithm [43]

The Luhn algorithm or Luhn formula is also known as the ——modulus 10¦ or ——mod 10¦ algorithm. This algorithm was created by IBM scientist Hans Peter Luhn [43].

This algorithm is used to check the validity of the credit card number.

8.3.4.1 Steps of Luhn Algorithm

- **Calculate the check digit:**

  The check digit is calculated by performing the sum of those digits which are non-check, thereafter computing 9 times that value modulo 10.

Step 1:

Starting from the check digit (rightmost digit), double the value of every second digit by moving from right to left.

Step 2:

If doubling of a number is greater than 9, then add up the digits to get a single digit.

Step 3:

Add up all the digits.

Step 4:

Multiply by 9.

Step 5:

The last digit is the check digit.

- **Check digit validation:**

  Step 1:

  Starting from the check digit (rightmost digit), double the value of every second digit by moving from right to left.

  Step 2:

  If doubling of a number is greater than 9, then add up the digits to get a single digit.

  Step 3:

  Add up all the digits.

  Step 4:

  If the total is divisible by 10, then the number is valid, otherwise the number is invalid.

8.3.4.2 Example of Luhn Algorithm

Consider the sample number ——**42997391786‖.**

1. **Double every second digit from right**

   $8 \times 2 = 16$

   $1 \times 2 = 2$

   $3 \times 2 = 6$

   $9 \times 2 = 18$

   $2 \times 2 = 4$

2. Sum digits of each multiplication 7, 2, 6, 9, 4

3. Add all digits

$$4 + 9 + 7 + 9 + 7 + 6 + 7 + 2 + 6 + 9 + 4 = 70$$

4. Calculate sum modulo 10

70 od 10 is equal to 0, which means that 42997391786 is a valid number as it passes the Luhn test.

8.3.4.3 Algorithm Used

The following algorithms describe the procedure of validating the user information.

a Algorithm/Pseudo Code 1: Validating User Data Notation

```
if(pnum.length!=9 || isNaN(pnum))
{
alert("policy number must be of 9 digits");
ret=false;
}
else if (tamt.length<1 || isNaN(tamt))
{
alert("amount must be integer value");
ret=false;
}
else if (cnum.length!=16 || isNaN(cnum))
{
alert("card number must be of 16 digits");
ret=false;
}
else if (cvv==null || cvv=="" || cvv.length!=3 || isNaN(cvv))
{
alert("cvv number must be of 3 digits");
ret=false;
}
else if (date==null || date=="")
{ alert("Please select valid date");
ret=false;
}
```

8.4 SIMULATION PERFORMED

8.4.1 Simulation Environment

The implementation uses the following tools and techniques:

- Google App Engine
- Eclipse Java 2018–2019

- Java Development Kit (JDK 1.7)
- Microsoft Windows 8.1

8.4.2 Google App Engine

Google App Engine is a platform that facilitate multiple users to design, develop and host web applications on Google's infrastructure. It was initially released in 2008 as a beta version. It was presented as an example of SaaS and PaaS cloud computing solutions, as it can easily be used by website designers and web application developers to create their applications using a large variety of tools provided by Google.

8.5 IMPLEMENTATION/EXECUTION

Application developed on Eclipse: An application of Mobile Recharge was developed using Eclipse which is deployed on Google Cloud Platform using server (Figure 8.1)

Application deployed on Google Cloud Platform: Once the application is developed on Eclipse it is connected to the server. The application runs on Eclipse when the server is running. Using the cloud shell, the server is connected from Eclipse to the cloud and the application is deployed on the GCP (Google Cloud Platform) (Figure 8.2).

Application running on server: After deployment on the GCP, the application runs on the server. It checks whether requests to the application are successful or not (Figure 8.3).

The result shows that the mobile has been recharged and encrypted data are created along with it (Figure 8.4).

The reader can use the following code for deployment at their end.

```
<!DOCTYPE html>
<html>
<head>
<meta charset="UTF-8">
<title>MOBILE RECHARGE PAYMENT</title>

<style>
*, *:before, *:after {
 -moz-box-sizing: border-box;
 -webkit-box-sizing: border-box;
box-sizing: border-box;
}

body {
font-family: 'Nunito', sans-serif;
color: #384047;
}

form {
```

```css
max-width: 500px;
margin: 10px auto;
padding: 10px 20px;
background: #f0f0f5;
border-radius: 8px;
float: left;
}

h2 {
margin: 0 0 30px 0;
text-align: center;
}

input[type="text"],
input[type="date"],
input[type="datetime"],
input[type="number"],
input[type="month"]
{
background: rgba(255,255,255,0.1);
border: none;
font-size: 16px;
height: 10px;
margin: 0;
outline: 0;
padding: 15px;
width: 100%;
background-color: #ffffff;
color: black;
box-shadow: 0 1px 0 rgba(0,0,0,0.03) inset;
margin-bottom: 10px;
}

button {
padding: 19px 39px 18px 39px; color:
#FFF;
background-color: #778899;
font-size: 18px;
text-align: center;
font-style: normal;
border-radius: 5px;
width: 100%;
border: 1px solid #3ac162;
border-width: 1px 1px 2px;
box-shadow: 0 -1px 0 rgba(255,255,255,0.1) inset;
margin-bottom: 10px;
font-weight: bold;
}

fieldset {
margin-bottom: 20px;
```

```css
border: none;
}

legend {
font-size: 1.4em;
margin-bottom: 10px;
font-weight: bold;
}
label { display:
block;
margin-bottom: 8px;
color: black;
font-weight: bold;
font-size: large;
}

label.light {
font-weight: 300;
display: inline;
}

.number {
background-color:#778899;
color: #fff;
height: 30px;
width: 30px;
display: inline-block;
font-size: 0.8em;
margin-right: 4px;
line-height: 30px;
text-align: center;
text-shadow: 0 1px 0 rgba(255,255,255,0.2);
border-radius: 100%;
}

.col-half {
padding-right: 10px;
float: left;
width: 50%;
}
.col-half:last-of-type {
padding-right: 0;
}

@media screen and (min-width: 480px) {

form {
max-width: 680px;
}
```

```
}
</style>

</head>
<body>
<form name="card_info" method="post" onSubmit="return
validate();" action="dictionaryelec">

<h2>MOBILE RECHARGE </h2>

<fieldset>
<legend><span class="number">1</span>Bill Details</legend>
<label for="number">Mobile Number:</label>
<input type="text" id="pnum" name="pnum">


<label for="amount">Amount to be recharged:</label>
<input type="text" id="tamt" name="tamt">

<legend><span class="number">2</span>Payment Details</legend>
<label for="cardnum">Card Number:</label>
     <input type="text" id="cnum" name="cnum">

<div class="row">
<div class="col-half">
<div class="input-group">
<label for="cvv">CVV:</label>
<input type="password" id="cvv" name="cvv">

</div>
</div>
<div class="col-half">

<div class="input-group">
<label for="expirydate">Expiry Date:</label>

<input type="month" name="edate" id="edate" size="30" max="2027-
03" min="2017-03">

</div>
</div>
</div>

<div class="row">
<div class="col-half">

<div class="input-group">
<button type="submit">Submit</button>

</div>
</div>
<div class="col-half">
```

```
<div class="input-group">
<button type="reset">Cancel</button>

</div>
</div>
</div>
</fieldset>
<input type="hidden" name="ptext" id="ptext" size="70">
</form>

</body>
<script type="text/javascript">
function validate()
{
    varpnum=document.getElementById("pnum").value;

<html>
vartamt=document.getElementById("tamt").value;
varcnum=document.getElementById("cnum").
value;
varcvv=document.getElementById("cvv").value;    var date=document.
getElementById("edate").value;

 varfdate=date.substr(0, 4)+date.substr(5, 2);
    var ret=true;

    if (pnum.length!=10 || isNaN(pnum))
    {
    alert("mobile number must be of 10 digits");
        ret=false;
    }
    else if (tamt.length<1 || isNaN(tamt))
    {
    alert("amount must be integer value");
        ret=false;
    }
    else if (cnum.length!=16 || isNaN(cnum))
    {
    alert("card number must be of 16 digits");
        ret=false;
    }
    else if (cvv==null || cvv=="" || cvv.length!=3 || isNaN(cvv))
    {
    alert("cvv number must be of 3 digits");
        ret=false;
    }
    else if (date==null || date=="")
    {
    alert("Please select valid date");
```

```
    ret=false;
    }

    varptext=pnum+tamt+cnum+cvv+fdate;
    var result = "";
    for (var i = 0; i <ptext.length; i++) {
          var c = ptext.charCodeAt(i);
    result += String.fromCharCode(3+c);
document.getElementById("ptext").value=result;
          return ret;
}
</script>
</html document.getElementById("ptext").value=result;
          return ret;
}
</script>
</html document.getElementById("ptext").value=result;
          return ret;
}
</script>
```

Application deployed on Google Cloud Platform: Once the application is developed on Eclipse it is connected to the server. The application runs on Eclipse when the server is running. Using the cloud shell, the server is connected from Eclipse to the cloud and the application is deployed on the GCP (Google Cloud Platform) (Figure 8.2).

Application running on server: After deployment on the GCP, the application runs on the server. It checks whether requests to the application are successful or not (Figure 8.3). The code is given below.

```
Import ava.io.PrintWriter;
import javax.servlet.http.*;

@SuppressWarnings("serial")
public class DictionaryElecServlet extends HttpServlet
{
    public void doPost(HttpServletRequestreq,
HttpServletResponseresp) throws IOException
{
    longstartTime = System.nanoTime();
    resp.setContentType("text/html");
    PrintWriter out=resp.getWriter();
    String ptext=req.getParameter("ptext");
          out.println("<body bgcolor=\"#F8F8FF\">");
out.println("<font color=\"red\"><h4>"+ptext+"</h4></font>");
longendTime = System.nanoTime();
longtotalTime = endTime - startTime;
out.println("Time taken is:"+totalTime);}
}
```

8.6 CONCLUSION AND FUTURE WORK

Cloud computing, together with the Internet of Things and big data, have gained much importance in the last few years. These areas are likely to flourish and see continuing growth in the coming years as companies move towards cloud-based tools and technologies, and adopt intelligent business rules, frameworks and policies. This chapter has explained various applications, services, deployment methods, uses, tools and limitations related to cloud computing. A brief code for design and development of Mobile Recharge Application and its successful deployment on the Google Cloud Platform has also been explained. Future plans include exploring and implementing the framework presented for the development of an optimized scheduling algorithm for PaaS-based applications in a secure cloud environment using symmetric and asymmetric techniques on cloud computing tools.

REFERENCES

[1] How Cloud Computing Works by Jonathan Strickland [https://computer.howstuffworks.com/cloud-computing/cloud-computing2.htm. Accessed on 24 July 2020]

[2] HCL Technology Q&A [https://www.hcltech.com/technology-qa/what-cloud-computing-what-are-services-offered. Accessed on 24 July 2020]

[3] Cloud Computing Services [https://india.emc.com/corporate/glossary/cloud-computing-services.htm. Accessed on 24 July 2020]

[4] *Cloud Computing: A better way of computing* [https://www.thinksys.com/services/cloud/ Accessed on 24 July 2020]

[5] *What is cloud computing? A beginner's guide* [https://azure.microsoft.com/en-in/overview/what-is-cloud-computing/ Accessed on 24 July 2020]

[6] Cloud computing [https://en.wikipedia.org/wiki/Cloud_computing. [Accessed on 24 July 2020]

[7] 75 Open Source Cloud Computing Apps [https://www.datamation.com/cloud-computing/75-open-source-cloud-computing-apps-1.html. [Accessed on 24 July 2020]

[8] Top 10 cloud tools [https://www.networkworld.com/article/2164791/cloud-computing/top-10-cloud-tools.html. [Accessed on 24 July 2020].

[9] Qubole Resources [https://www.qubole.com/resources. Accessed on 24 July 2020]

[10] *Cognitive Top 7 most common uses of cloud computing* [https://www.ibm.com/blogs/cloud-computing/2014/02/06/top-7-most-common-uses-of-cloud-computing/. Accessed on 24 July 2020].

[11] *Cloud Computing Applications* [https://www.w3schools.in/cloud-computing/cloud-computing-applications/. Accessed on 26 July 2020]

[12] *Application of Cloud Computing in Various Sectors Information Technology Essay* [https://www.uniassignment.com/essay-samples/information-technology/application-of-cloud-computing-in-various-sectors-information-technology-essay.phped. Accessed on 26 July 2020]

[13] *Cryptography in the Cloud: Securing Cloud Data with Encryption by Nate Lord* [https://digital-guardian.com/blog/cryptography-cloud-securing-cloud-data-encryption. Accessed on 26 July 2020]

[14] *Crypto cloud computing* [https://en.m.wikipedia.org/wiki/Crypto_cloud_computing. Accessed on 26 July 2020]

[15] *Designing business information systems: Apps, websites, and more* [https://2012books.lardbucket.org/books/designing-business-information-systems-apps-websites-and-more/. Accessed on 26 July 2020]

[16] *Load balancing algorithms and techniques* [https://kemptechnologies.com/in/load-balancer/load-balancing-algorithms-techniques/. Accessed on 26 July 2020]

[17] *Different-types-of-load-balancing-algorithm-techniques* [http://apachebooster.com/kb/different-types-of-load-balancing-algorithm-techniques/. Accessed on 26 July 2020]

[18] *Resource scheduling* [https://www.techopedia.com/definition/30464/resource-scheduling. Accessed on 26 July 2020]

[19] *Resource scheduling algorithm by Shilpa Damor* [https://www.slideshare.net/mobile/shilpadamor9/resource-scheduling-algorithm. Accessed on 26 July 2020]

[20] K. Li, G. Xu, G. Zhao, Y. Dong, and D. Wang. Cloud task scheduling based on load balancing ant colony optimization. In *2011 Sixth Annual ChinaGrid Conference*, Liaoning, China, August 2011, pp. 3–9. IEEE.

[21] M. S. Rana and N. Jaisankar. Comparison of probabilistic optimization algorithms for resource scheduling in cloud computing environment. *International Journal of Engineering and Technology*, 5(2):1419–1427, 2013.

[22] S. Sangwan and S. Sangwan. An effective approach on scheduling algorithm in cloud computing. *International Journal of Computer Science and Mobile Computation*, 3(6):19–23, 2014.

[23] R. J. Priyadarsini and L. Arockiam. A framework to optimize task scheduling in cloud environment. *International Journal of Computer Science and Information Technologies*, 5(6):7060–7062, 2014.

[24] B. Santhosh, D. H. Manjaiah, and L. P. Suresh. A survey of various scheduling algorithms in cloud environment. In *International Conference on Emerging Technological Trends (ICETT)*, Kollam, India, October 2016, pp. 1–5. IEEE.

[25] S. Singh, Y. S. Jeong, and J. H. Park. A survey on cloud computing security: issues, threats, and solutions. *Journal of Network and Computer Applications*, 75:200–222, 2016.

[26] S. P. Praveen, K. T. Rao, and B. Janakiramaiah. Effective allocation of resources and task scheduling in cloud environment using social group optimization. *Arabian Journal for Science and Engineering*, 43:4265–4272, 2017.

[27] Y. Yu, A. Miyaji, M. H. Au, and W. Susilo. Cloud computing security and privacy: standards and regulations. *Computer Standards & Interfaces*, 54:1–2, 2017.

[28] A. Kaur, B. Kaur, and D. Singh. Optimization techniques for resource provisioning and load balancing in cloud environment: a review. *International Journal of Information Engineering and Electronic Business*, 9(1):28, 2017.

[29] G. Ramachandra, M. Iftikhar, and F. A. Khan. A comprehensive survey on security in cloud computing. *Procedia Computer Science*, 110:465–472, 2017.

[30] C. A. B. De Carvalho, R. M. de Castro Andrade, M. F. de Castro, E. F. Coutinho, and N. Agoulmine. State of the art and challenges of security SLA for cloud computing. *Computers & Electrical Engineering*, 59:141–152, 2017.

[31] N. Hamdy, A. Elsayed, N. ElHaggar, and M. Mostafa-Sami, Resource allocation strategies in cloud computing: overview. *International Journal of Computer Applications*, 975:8887, 2017.

[32] I. Giannakopoulos, I. Konstantinou, D. Tsoumakos, and N. Koziris. Cloud application deployment with transient failure recovery. *Journal of Cloud Computing*, 7(1):11, 2018.

[33] S. N. S. Kamil and N. Thomas. Investigating the cost of transfer delay on the performance of security in cloud computing. *Electronic Notes in Theoretical Computer Science*, 337:105–117, 2018.

[34] C. Stergiou, K. E. Psannis, B. B. Gupta, and Y. Ishibashi. Security, privacy & efficiency of sustainable cloud computing for Big Data & IoT. *Sustainable Computing: Informatics and Systems*, 19:174–184, 2018.

[35] T. S. Alnusairi, A. A. Shahin, and Y. Daadaa. Binary PSOGSA for load balancing task scheduling in cloud environment. *International Journal of Advanced Computer Science and Applications*, 9(5):255–264, 2018.

[36] V. Vishal and R. Johari. SOAiCE: simulation of attacks in cloud computing environment. In *2018 8th International Conference on Cloud Computing, Data Science & Engineering (Confluence)*, January 2018, pp. 14–15. IEEE.

[37] S. Gupta, R. Johari, P. Garg, and K. Gupta. C³T: Cloud based Cyclic Cryptographic Technique and it's comparative analysis with classical cipher techniques. In *2018 5th International Conference on Signal Processing and Integrated Networks (SPIN)*, New Delhi, India, February 2018, pp. 332–337. IEEE.

[38] S. Gupta and R. Johari. UID C: Cloud based UID application. In *2017 7th International Conference on Cloud Computing, Data Science & Engineering-Confluence*, Noida, India, January 2017, pp. 319–324. IEEE.

[39] P. Grover and R. Johari. BCD: BigData, cloud computing and distributed computing. In *2015 Global Conference on Communication Technologies (GCCT)*, Thuckalay, India, April 2015, pp. 772–776. IEEE.

[40] P. S. Yoganandani, R. Johari, K. Krishna, R. Kumar, and S. Maurya. Clearing the clouds on cloud computing: survey paper. *International Journal of Recent Development in Engineering and Technology*, 1:117–121, 2014.

[41] S. Koushal and R. Johri. Cloud simulation environment and application load monitoring. In *2013 International Conference on Machine Intelligence and Research Advancement (ICMIRA)*, Katra, JK, India, December 2013, pp. 554–558. IEEE.

[42] Credit card [https://en.wikipedia.org/wiki/Credit_card. Accessed on 26 July 2020].

[43] Luhn algorithm. [https://en.wikipedia.org/wiki/Luhn_algorithm. Accessed on 26 July 2020].

CHAPTER 9

UAV Environment in FANET: An Overview

Dipta Datta, Kavita Dhull, and Sahil Verma

CONTENTS

9.1 INTRODUCTION

Unmanned aerial vehicles (UAVs) have today become very popular for their wide communication coverage and easy installation [1, 2]. They are used in mission-based operations such as rescue, goods delivery, military missions and patrols. Normally UAVs can be used as sensors, internal hops and base stations [3, 4]. However, UAVs require a reliable network when flying, as well as safety, coordination, communication and energy efficiency [2]. Effective routing techniques and protocols are required to maintain security and power management cost-effectively. Among the challenging issues faced by UAVs are security in the server, connectivity and energy drain at high speed. Moreover, unpredictable movements and non-uniform distribution frequently change the topology over the network [5]. Designing any routing protocol for a Flying Ad-Hoc Network (FANET) is hence a very difficult task. Cloud computing plays a very important role in FANET. Secured cloud servers are used for different missions to make the information safe. Cloud applications are installed in specific nodes to enable proper communication.

Although UAVs can change direction and move very fast, their mobility delays data transmission and slows down maintenance [1]. Survivability, reliability and scalability of UAVs makes the network highly convenient [6, 7]. There are other subclasses of ad-hoc network besides FANET: mobile ad-hoc network (MANET), vehicular ad-hoc network (VANET), ship ad-hoc network (SANET), ad-hoc robot network (RANET) and wireless sensor network (WSN) [8]. FANET vehicles are very fast (achieving speeds of up to 460 km/h) and are used for data gathering in dangerous or normally inaccessible places. As FANET is used in many secret and military missions, security issues and energy efficiency are very important for maintaining connectivity with the control station.

9.2 RELATED WORK

As FANET is becoming a major part of the aerial environment, there have been an increasing number of studies in this sector producing different solutions and techniques. Different topologies, mobility models, routing techniques, routing protocols and so on have been addressed.

Mahmud and Cho [1] discuss green UAVs and introduce the energy-efficient "hello message" system which saves about 25 per cent of energy. Oubbati et al. [2] describe most of the mobility models, routing techniques and protocols of FANET. Park et al. [3] provide an algorithm for a scenario where UAVs can reconstruct ad-hoc networks damaged by natural disaster. Gong et al. [4] tried to minimize the flight time of UAVs flying in a straight line with non-overlapping data collection intervals using fixed given energy. Wang et al. [5] describe an ultra-dense network (UDN) that supports UAVs. Wen et al. [8] propose an adaptive distributed routing method to minimize the complexity of routing algorithms in real time. They tried to maximize the utility at end-to-end delay within the proper threshold [8]. Khan et al. [9] introduced two optimization algorithms named glowworm swarm optimization (GSO) and krill herd (KH) to minimize FANET energy consumption. They also used ant colony optimization to build the clustering scheme. Tareque et al. [10] categorized FANET routing protocols and compared them with other ad-hoc networks. Vasiliev et al. [11] considered source, destination and relay as nodes and investigated throughput efficiencies of relaying algorithms. Kerrache et al. [12] established a trust-based energy-efficient FANET monitoring technique, seeking to establish the trust of peers and distributing monitoring periods among them. This proposal gave high detection rates and more energy efficiency [12]. Gankhuyag et al. proposed a scheme that combines both omnidirectional and directional transmission with unicasting and geocasting routing [13]. Here, path re-establishment and service disruption are reduced to increase the path lifetime and successful packet transmission [13]. Sobhy et al. [14] investigated UAV cloud computing using throughput for both Windows and Linux. They also introduced a cloud computing infrastructure providing better cloud storage and other facilities for FANET [15]. Radu et al. [16] used GPS and video-enabled drones for data gathering, and the cloud system to provide security. Savita et al. [17] proposed a decentralized cloud system with specific sensors and base station. Different nodes are connected to the stations to extend the network coverage. Mahmoud et al. [18] explained different opportunities and challenges of the cloud in FANET. They also proposed a cloud system for UAVs to minimize deployment cost and time, and a FANET architecture in which UAVs would act as servers or parts of the cloud system [19]. Tiwari et al. [20] proposed an architecture for FANET that could monitor flight times. They used the cloud system to build the architecture for the air navigation system.

9.3 FANET OVERVIFW

Although FANET is similar to MANET [2], some modifications are required. Flying constraints include mobility, deployment, cost-effectiveness, coverage area, QoS, energy and security. The cloud system can minimize the constraints with the help of different types of techniques and models.

9.3.1 Categories of UAVs

In FANET, UAVs can be categorized by altitude (Figure 9.1). High-altitude UAVs (above 20 km from the ground) are known as HAUs, medium-altitude UAVs (up to 11 km) are MAUs, and low-altitude UAVs (just a few km) are LAUs [21].

Normally HAUs are satellites, airships and hot-air balloons connected with the FANET satellite network. Airships are MAUs and can move faster than the ground nodes. Drones and copters are LAUs, connected with the ground network to communicate with other devices and networks.

According to the node speed, UAVs can be categorized into two types: (i) rotary wing (RW), and (ii) fixed wing (FW). The mobility models of both types are the same; both can be deployed in 2D or 3D space. However, at up to 100 m/s, FWs can fly very much faster than RWs (up to 30 m/s), which is why FWs are more popular and can manage both medium and low altitudes.

9.3.2 UAV Communication

UAVs are normally used in restricted areas not easily accessible to people. Sudden disconnection, packet loss fixed fragmentation are the major issues. The FANET communication

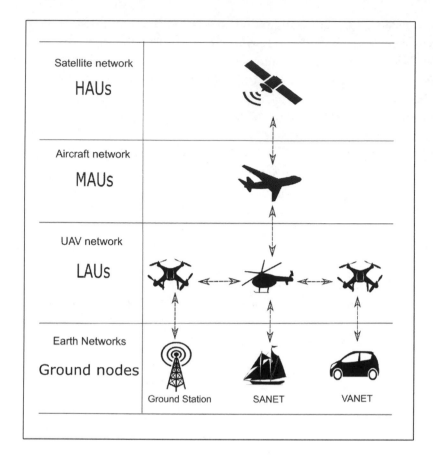

FIGURE 9.1   Categories of UAV.

process is divided into three categories: (i) UAV-to-UAV communication (U2U), (ii) UAV-to-ground communication (U2G) and (iii) satellite communication (SATCOM) [2].

U2U communication is used in most missions and can safely exchange packets with every UAV. All UAVs share data with each other, so that if any UAV becomes corrupted or is disconnected, other UAVs have all the data. As a result, packet loss is minimized. If there is no obstacle, the line of sight (LoS) rises high in U2U communication [22].

U2G communication is needed to control the network and collect the data. A few UAVs are connected to the control station for transfer of all data. Generally, specific UAVs can communicate with ground base stations (GBSs) to minimize network congestion. When UAVs fly at high altitude, the LoS rises in the U2G connection, and when they fly at low altitudes, due to different obstacles, LoS cannot be ensured in the U2G connection [23]. This communication uses a cloud system to provide secure transmission.

Sometimes UAVs are deployed in places where a GBS cannot be installed. Here, satellites are a suitable communication option. Satellites can provide safe and continuous connectivity. Satellites help the UAVs to exchange data, and satellites send the data to verified radio stations. SATCOM is therefore useful for gathering data from complex areas and represents a secure path for the exchange of crucial data, but it is not a cost-effective option.

9.4 ROUTING TECHNIQUES

FANET uses different routing techniques for complete communication. Though different mobility models make the design complicated, different routing techniques are used to complete communication and make it safe. Routing techniques, based on the situation and the environment, are used to reduce packet loss and give better outcomes. This section describes the most popular and frequently used routing techniques that maintain communication and help to control the network (for an overview see Figure 9.2).

9.4.1 Clustering

It is often the case that many UAVs are deployed for a particular mission, causing increased density of FANET. In this situation, different clusters are made, each consisting of a small

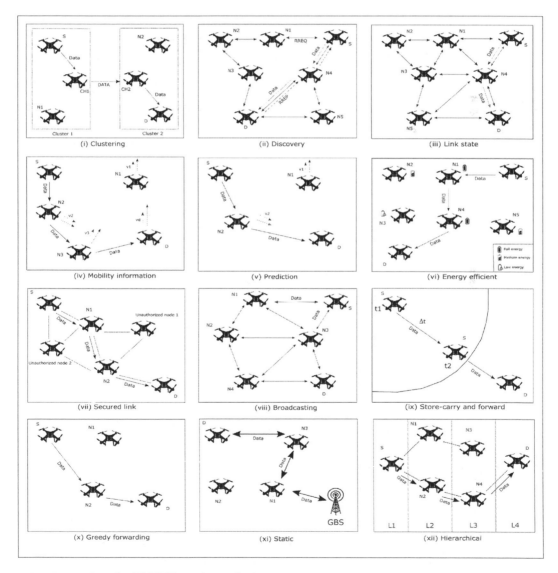

FIGURE 9.2   Popular FANET routing techniques.

number of UAVs. These clusters are controlled by cluster heads to which UAVs are connected. Figure 9.2 (i) illustrates a scenario where source UAV (S) wants to send data to destination UAV (D). First, the source sends data to the cluster head (ch1) of cluster 1; then, ch1 sends the data to the other cluster head (ch2) of cluster 2; and finally, ch2 transfers the data to the destination. Though this technique divides the network into different clusters, it cannot be used in a low-density network.

9.4.2 Discovery

Sometimes in FANET, the location of the destination is not known, and FANET uses a discovery technique to find it. Normally, the source sends a route request (RREQ) to all connected nodes. When the destination receives the RREQ, it replies with a route repeat (RREP). After receiving the RREP, the source plots the path to the destination and sends the data. For instance, source UAV (S) sends RREQ to nearby nodes, and when destination UAV (D) receives the request, it replies with an RREP (see Figure 9.2(ii)). Then the source uses the path that has been discovered to send packets.

9.4.3 Link State

In this technique, all UAVs receive link-state information from all nodes and share the information with other nearby UAVs. The source then uses the shortest path from the link-state information to transfer the data to the destination (see Figure 9.2(iii)). As all link-state information needs to be shared with all nodes, this technique has low overheads.

9.4.4 Mobility Information

FANET uses the position, speed and velocity parameters of the UAVs to confirm the next relay. Using the parameters, UAVs can select the correct path to send data to the destination. As node 2 UAV (N2) and node 3 UAV (N3) are directed towards the destination UAV (D), the source UAV (S) use the path (S-N2-N3-D) to transfer the data (see Figure 9.2(iv)). This topology needs frequent "hello" packets within short fixed intervals. Every UAV sends a hello packet to others to confirm the establishment of the network. Though this technique improves the connectivity of FANET, it cannot be used in different fragmented networks.

9.4.5 Prediction

Sometimes the speed and direction of a UAV confirms its next position. For instance, source UAV (S) uses movement details about the nearby UAVs to confirm the best path and to send data to the destination UAV (D) (see Figure 9.2(v)). This topology needs detailed information about nearby nodes, but the technique minimizes node disconnections and requires high density to perform communications.

9.4.6 Energy Efficiency

FANET requires balanced energy efficiency if it is to ignore any packet loss due to sudden disconnection. Normally UAVs with low energy are ignored in any communication to minimize packet loss. Source UAV (S) selects the path that has full or minimum energy to complete the transmission (see Figure 9.2(vi)). Though the technique separates the nodes with low energy, it does not care about the connectivity of the nodes.

9.4.7 Secured Link

In FANET, different security techniques are used to secure communications, find unauthorized UAVs and ignore them in the transmission process. When any unauthorized UAV connects to the network, different security algorithms or protocols detect the unauthorized node transfer and ignore them, transferring the data through the authorized UAVs (see Figure 9.2(vii)). Various complex security algorithms or protocols are required to detect malicious nodes.

9.4.8 Broadcasting

In this technique, all the UAVs get data from the source. The data packet is shared with all nodes within the network. Source UAV (S) shares data with the nearby nodes, which share the data with other nodes to make the data available to all nodes within the network (see Figure 9.2(viii)). With the data broadcast in the network, high overhead and congestion problems arise. But this technique ensures the highest number of successful transmissions [2].

9.4.9 Store and Carry Forward

When the destination is not in the source's coverage area, the source carries the data until it meets the destination. For instance, destination UAV (D) is not in the range of source UAV (S) (see Figure 9.2(ix)). UAV (S) carries the data during Δt ($\Delta t = t2 - t1$), and when the UAV (D) comes in the range, (S) forwards the data to the destination.

9.4.10 Greedy Forwarding

In this technique, the source forwards the data to the node that is closest to the destination, reducing unnecessary relays. In Figure 9.2(x), N2 UAV is closest to the destination UAV (D). So, source UAV (S) forwards the data to N2 and then N2 forwards data to (D). Though the technique reduces the number of nodes and delays, it can fail at local optimums [2].

9.4.11 Static Technique

In this technique, the GBS sets up a routing table to communicate with the UAVs. According to the table, data transmission gets done using the fixed routing path (see Figure 9.2(xi). For different types of topologies, this technique does not work properly.

9.4.12 Hierarchical Technique

This technique divides the network into different levels. Each level has its own root UAVs. Communication is established by the root UAVs. Root UAVs connect the levels and help to complete the transmission (see Figure 9.2(xii)). The disadvantage of the technique is that it does not work for high mobility.

9.5 CLOUD COMPUTING IN FANET

The Internet of Things (IoT) and the cloud system have brought revolutionary changes in the FANET system, reducing packet loss, delays and the problems of limited resource. Cloud computing is a very important part of FANET; it gathers data for the server and provides security for the network. Cloud computing provides effective organization of the network, and arranges the use of computing power, infrastructure and applications through the network. To avoid congestion and overhead, the cloud computing system tries to optimize the limited resources according to the demand of the network. Cloud computing is an automated system that can connect different clusters across a big area and transfers data to the main controller. Normally, UAVs gather data from different fields (agriculture, VANET, SANET, mountain areas, critical mission-related areas) and transfer them immediately to the cloud system (Figure 9.3). Then the cloud safely transfers the data to the controller station through the secured gateway path.

Cloud computing applications can be used for different purposes. UAVs can gather important data from agricultural areas, mountains or oceans that can be used to provide solutions in these areas. Military missions also use UAVs to obtain critical data through a cloud system. Users can access data using military networks instead of the internet through the cloud system [24]. However, the cloud computing system also needs a proper security model to provide safe communication.

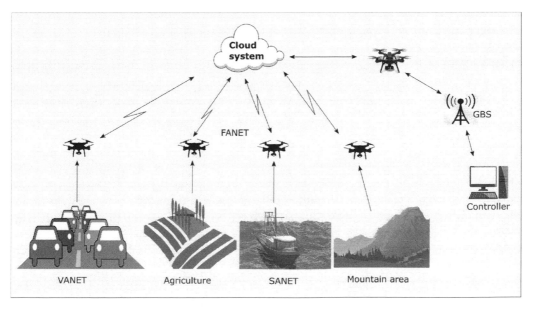

FIGURE 9.3   Cloud computing in FANET.

9.6 MOBILITY MODELS

The mobility model is another challenge for FANET. Different models have their own strategies as regards movement, speed and direction of UAVs. In time-based mobility models, UAVs change speed and direction at fixed intervals. In random-based mobility models, speed and direction can change randomly. In path-based mobility models the UAVs calculate their next position. In group-based mobility models, they move together to complete the mission. Topology-based mobility models are used to move within fixed locations. A simple overview of all the mobility models is given in Table 9.1.

TABLE 9.1   Overview of FANET Mobility Models

Type of Model	Model	Movement Topology
Random-based	RW (Random Walk) [25]	Randomly move to boundaries at fixed time
	RWP (Random Waypoint) [26]	Randomly add pause between two changes of direction
	RD (Random Direction) [27]	Move toward the boundary and pause a while before the next direction
	MG (Manhattan Grid) [28]	Randomly move within the boundary ranges
Time-based	BSA (Boundless Simulation Area) [29]	When it reaches the boundary, it does not bounce on the border and appears to the opposite boundary.
	GM (Gauss-Markov) [30]	Discover the next movement, according to the previous one.
	ST (Smooth Turn) [31]	Randomly selects different points and, according to the points, turns with radius.
	3WR (Three-Way Random) [32]	Randomly selects left turn, right turn and straight ahead.
Path-based	SRCM (Semi-Random Circular Movement) [33]	Moves around a fixed round area.
	PPRZM (Paparazzi Mobility Model) [34]	Uses five types of motion according to the situation.
	MT (Multi-Tyre) [35]	Supports multiple models at a time.
Group-based	ECR (Exponential Correlated Random) [32]	Uses motion information for possible movements.
	PRS (Purse Mobility Model) [36]	The group of nodes moves together to reach the destination.
	PSMM (Particle Swarm Mobility Model) [37]	Finds out the movement direction of other nodes with the help of a reference point.
Topology-based	DPR (Distributed Pheromone Repel) [32]	Uses pheromones to collect the movements of other nodes.
	SDPC (Self-Deployable Point Coverage) [38]	Spreads coverage for the ground networks.

9.7 ROUTING PROTOCOLS

Different constraints create serious FANET design challenges. Highly dynamic topology, security, energy consumption, recovery of broken links, and resource fixes are the major constraints, and these constraints increase the challenges of building a perfect FANET system. Different routing protocols have been established to provide solutions. These protocols can be divided into different categories according to the network situation (see Figure 9.4) [2]. Each type has different protocols, providing solutions for different constraints.

9.7.1 Energy-Based Routing Protocols

There are many routing protocols aiming to balance energy in FANET. These protocols try to consider issues other than energy efficiency. Different hello interval techniques are used to minimize energy consumption. Many energy-based routing protocols are used to increase UAV lifetime.

CBLADSR (Cluster-based location-aided dynamic source routing) [39] is one of the energy-based routing protocols. It uses the concept of cluster groups and the dynamic source routing strategy [40]. This protocol also supports geographical routing. The clustered groups use the discovery technique to find out information from nearby clusters. The source transfers the data through nearby UAVs that have enough energy to complete the task, ignoring UAVs with low energy resources. The cluster head transfers the data from the source cluster to the destination cluster. The destination receives the data from the cluster head through other UAVs with minimum energy to complete the transmission. This routing protocol generally supports nodes containing high energy and requires a high-density network. Though the successful data transmission rate is high, this protocol has high delay rates and overhead.

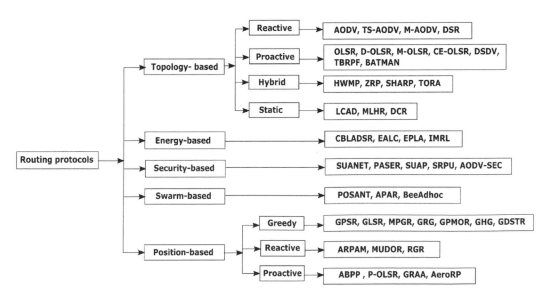

FIGURE 9.4   FANET routing protocols.

Energy-aware link-based clustering (EALC) [41] is another energy-based routing protocol that uses the concept of clustering. In EALC, every cluster collects the information of other clusters with a high lifetime. This protocol can decrease packet loss by transferring data only through high-powered clusters. The K-mean algorithm is used in this clustering technique. This routing protocol is used when the source tries to send data to the ground station. The source selects nearby clusters with high energy, and the cluster heads transmit the data through the UAVs inside the clusters. This protocol reduces packet loss, but at the expense of high energy consumption.

The energy-efficient packet load algorithm (EPLA) [42] is one of the energy-based routing protocols that uses a transmission schedule to reduce energy consumption. New UAVs can be added to the network to transmit the data to the ground station, but their energy levels need to be high. This protocol selects UAVs with high energy to find the route to the destination. It increases network lifetime and reduces the error rate, but does not consider any link failure.

IMRL (Localization and Energy-efficient Data Routing for UAVs) [43] uses the geographical position of other UAVs to transfer data. The source sends data to the cluster head, which finds a cluster whose head is very near and that has high energy. The data are transferred to the ground station through the shortest path using shortest-path algorithms. This protocol increases the lifetime of the network, but it cannot support a high-mobility network. Table 9.2 represents the advantages and disadvantages of energy-based routing protocols.

9.7.2 Security-Based Routing Protocols

Security and privacy are the most important part of any network, so different security techniques are used to secure networks. In FANET, security protocols are used to detect unauthorized malicious nodes and to prevent attacks. These protocols only allow authorized nodes to take part in data transmission.

SUANET (Secure UAV Ad-hoc Network) [44] is a security-based routing protocol that verifies nodes authorized for communication. This protocol generates keys to verify UAVs. The routing process verifies authorized UAVs with key authentication and finds the best

TABLE 9.2   Advantages and Disadvantages of Energy-Based Routing Protocols

	Protocols	Advantages	Disadvantages
Energy-based	CBLADSR	High complete transmission ratio.	High delay and overhead.
	EALC	Minimizes packet loss.	High energy consumption in each cluster.
	EPLA	Establishes balanced energy among UAVs.	Fails to consider link damages.
	IMRL	Increases the lifetime of the network.	It does not work in high mobility.

and shortest path to the destination. The algorithms used in this protocol are highly complex, but the protocol provides high network security.

PASER (Position-aware, Secure and Efficient Mesh Routing) [45] is a secure-based routing protocol that uses cryptographic algorithms to provide security in FANET. PASER can easily detect malicious UAVs within the network and blocks connections with them. This protocol has high overhead and delay.

SUAP (Secure UAV Ad-hoc Routing Protocol) [46] is another secure-based routing protocol that uses message authentication algorithms to detect and prevent attacks. In this protocol, encrypted packets are transmitted, and digital signatures are used to detect wormhole attacks. Authorized UAVs use the source's public key to decrypt packets. To measure the distance, this protocol counts the hops. The source sends RREQ within the network to find the best path to the destination, and the destination RREP confirms the path with RREP. Then SUAP verifies the path and confirms it as secure.

SRPU (Secure Routing Protocol for UAVs) [47] is another security-based routing protocol, which provides more secure data transmission. It repeatedly calculates the distance that packets transit and their departure times. If any attack tries to change the packets, comparing the distance and departure times, SRPU confirms the attacks.

AODV-SEC (Ad-hoc On-demand Distance Vector Secure) [48] assigns a public key, which is used by the network to find the best path to reach the destination. At the same time, it detects malicious nodes and ignores them while transferring data. In AODV-SEC, RREQ-ACK is sent from the destination UAV. This RREQ-ACK is used to ignore unnecessary and malicious RREP from unauthorized UAVs. When RREP is delivered to the source, packet verification confirms the validation of the request. Only when both packet requests are verified by the system can the transmission process start.

Table 9.3 represents the advantages and disadvantages of security-based routing protocols.

TABLE 9.3   Advantages and Disadvantages of Security-Based Routing Protocols

	Protocols	Advantages	Disadvantages
Security-based	SUANET	Improves security among the links.	Does not consider stability
	PASER	Improves security with scalability.	High delay and overhead
	SUAP	Ensures secure path.	Does not work in high mobility
	SRPU	Improves security and ignores attacks.	High overhead.
	AODV-SEC	Improves secure path finding.	Applies difficult algorithms and processes

9.8 CHALLENGES AND SOLUTIONS

Although FANET has brought revolutionary changes to modern networks, it has many constraints. High mobility, packet loss, random and high-speed movement, sudden disconnection, limited bandwidth, energy limitations and security breaches are the main challenges for FANET. As cloud computing is a very important part of FANET, the security of the cloud system is also a big issue in this field. All data transmissions are completed through the cloud to make the best use of limited resources. But if the cloud system is not secure, the whole network cannot provide proper security for the information. So, a proper security model needs to be installed for the cloud system.

Energy is a very important part of UAVs in FANET. But lack of energy may decrease the lifetime of the network. More battery development work, along with positioning recharge stations in common areas, is needed to reduce energy problems. Many researchers are working on hello message interval algorithms to increase battery life. Despite the many modern security techniques, FANET can be attacked by any malicious UAV. As FANET is normally controlled by the ground station, any air attack can occur in the network. UAVs are generally used in restricted areas. U2U communication enables data sharing and easy transfer to the destination, but U2U communication with regular traffic can be a challenge.

Many FANET networks use cryptographic algorithms to protect the network from attacks, but this is an energy-intensive process. In FANET, UAVs fully depend on each other, but if any malicious UAV enters the network, the entire network can be compromised. So more emphasis on detection and prevention systems is needed. Routing is another issue. High mobility is the major reason for failure to establish perfect routing techniques. No proposed technique can deal completely with all the issues, so an appropriate network design solution is needed.

9.9 CONCLUSION

UAVs play a very important role in the air today. The FANET design is very advanced, with its different applications, routing techniques, mobility models and routing protocols – all of which have their own characteristics, advantages and disadvantages. As a result, not all needs can be fulfilled easily. A high-density network consisting of many UAVs is needed, as they are mutually dependent.

The two most important topics in FANET are security and energy efficiency. Fixed energy is the common drawback of UAVs. With fixed energy, neither security nor data communication can work properly. Securing the network and controlling the nodes takes most of the energy. Complete design with security and energy efficiency can bring a revolutionary change in FANET.

This chapter has reviewed the FANET system in relation to cloud computing, and has explained different routing techniques and protocols. The challenges for FANET have been outlined and a number of cloud-based solutions described.

REFERENCES

[1] I. Mahmud and Y.-Z. Cho. Adaptive hello interval in FANET routing protocols for green UAVs. *IEEE Access*, 7:63004–63015, 2019.

[2] O. S. Oubbati, M. Atiquzzaman, P. Lorenz, Md. H. Tareque, and Md. S. Hossain. Routing in flying ad hoc networks: survey, constraints, and future challenges perspectives. *IEEE Access*, 7:81057–81105, 2019.

[3] S.-Y. Park, C. S. Shin, D. Jeong, and H. J. Lee. DroneNetX: network reconstruction through connectivity probing and relay deployment by multiple UAVs in ad hoc networks. *IEEE Transactions on Vehicular Technology*, 67(11):11192–11207, 2018.

[4] J. Gong, T.-H. Chang, C. Shen, and X. Chen. Flight time minimization of UAV for data collection over wireless sensor networks. *IEEE Journal on Selected Areas in Communications*, 36(9):1942–1954, 2018.

[5] H. Wang, G. Ding, F. Gao, J. Chen, J. Wang, and L. Wang. Power control in UAV-supported ultra dense networks: communications, caching, and energy transfer. *IEEE Communications Magazine*, 56(6):28–34, 2018.

[6] I. Jawhar, N. Mohamed, J. Al-Jaroodi, D. P. Agrawal, and S. Zhang. Communication and networking of UAV-based systems: classification and associated architectures. *Journal of Network and Computer Applications*, 84:93–108, 2017.

[7] L. Gupta, R. Jain, and G. Vaszkun. Survey of important issues in UAV communication networks. *IEEE Communications Surveys & Tutorials*, 18(2):1123–1152, 2015.

[8] S. Wen, C. Huang, X. Chen, J. Ma, N. Xiong, and Z. Li. Energy-efficient and delay-aware distributed routing with cooperative transmission for the Internet of Things. *Journal of Parallel and Distributed Computing*, 118:46–56, 2018.

[9] A. Khan, F. Aftab, and Z. Zhang. BICSF: bio-inspired clustering scheme for FANETs. *IEEE Access*, 7:31446–31456, 2019.

[10] Md. H. Tareque, Md. S. Hossain, and M. Atiquzzaman. On the routing in flying ad hoc networks. In *2015 Federated Conference on Computer Science and Information Systems (FedCSIS)*, Lodz, Poland, pp. 1–9. IEEE, 2015.

[11] D. S. Vasiliev and A. Abilov. Relaying algorithms with ARQ in flying ad hoc networks. In *2015 International Siberian Conference on Control and Communications (SIBCON)*, Omsk, Russia, pp. 1–5. IEEE, 2015.

[12] C. A. Kerrache, E. Barka, N. Lagraa, and A. Lakas. Reputation-aware energy-efficient solution for FANET monitoring. In *2017 10th IFIP Wireless and Mobile Networking Conference (WMNC)*, Valencia, Spain, pp. 1–6. IEEE, 2017.

[13] G. Gankhuyag, A. P. Shrestha, and S.-J. Yoo. Robust and reliable predictive routing strategy for flying ad-hoc networks. *IEEE Access*, 5:643–654, 2017.

[14] A. R. Sobhy, A. T. Khalil, M. M. Elfaham, and A. Hashad. UAV cloud operating system. In *MATEC Web of Conferences*, Chios, Greece, Vol. 188, p. 05011. EDP Sciences, 2018.

[15] A. R. Sobhy, M. M. Elfaham, and A. Hashad. Fanet cloud computing. *International Journal of Computer Science and Information Security*, 14(10):88, 2016.

[16] D. Radu, A. Cretu, B. Parrein, J. Yi, C. Avram, and A. Aştilean. Flying ad hoc network for emergency applications connected to a fog system. In *International Conference on Emerging Internetworking, Data & Web Technologies*, pp. 675–686. Springer, Cham, 2018.

[17] A. Savita, P. Sharma, and S. K. Tiwari. Cloud-based decentralized approach for FANET. In *2016 International Conference on Signal Processing, Communication, Power and Embedded System (SCOPES)*, Odisha, India, pp. 2017–2020. IEEE, 2016.

[18] S. Mahmoud and N. Mohamed. Collaborative UAVs cloud. In *2014 International Conference on Unmanned Aircraft Systems (ICUAS)*, Orlando, FL, pp. 365–373. IEEE, 2014.

[19] S. Mahmoud and N. Mohamed. Broker architecture for collaborative UAVs cloud computing. In *2015 International Conference on Collaboration Technologies and Systems (CTS)*, Atlanta, GA, pp. 212–219. IEEE, 2015.

[20] V. Tiwari, K. Sharma, and B. K. Chaurasia. FANET based flights monitoring simulation system over cloud. In *Advances in Optical Science and Engineering*, pp. 417–423. Springer, New Delhi, 2015.

[21] M. Alzenad, M. Z. Shakir, H. Yanikomeroglu, and M.-S. Alouini. FSO-based vertical backhaul/fronthaul framework for 5G+ wireless networks. *IEEE Communications Magazine*, 56(1):218–224, 2018.

[22] M. Mozaffari, W. Saad, M. Bennis, Y.-H. Nam, and M. Debbah. A tutorial on UAVs for wireless networks: applications, challenges, and open problems. *IEEE Communications Surveys & Tutorials*, 21(3):2334–2360, 2019.

[23] A. A. Khuwaja, Y. Chen, N. Zhao, M.-S. Alouini, and P. Dobbins. A survey of channel modeling for UAV communications. *IEEE Communications Surveys & Tutorials*, 20(4):2804–2821, 2018.

[24] A. R. Sobhy, A. T. Khalil, M. M. Elfaham, and A. Hashad. Proposed algorithms for UAV based cloud computing. *International Journal of Computer Science and Information Security (IJCSIS)*, 16(2), 2018. 122–128.

[25] K.-H. Chiang and N. Shenoy. A 2-D random-walk mobility model for location-management studies in wireless networks. *IEEE Transactions on Vehicular Technology*, 53(2):413–424, 2004.

[26] J. Yoon, M. Liu, and B. Noble. Random waypoint considered harmful. In *IEEE INFOCOM 2003. Twenty-Second Annual Joint Conference of the IEEE Computer and Communications Societies* (IEEE Cat. No. 03CH37428), San Francisco, CA, Vol. 2, pp. 1312–1321. IEEE, 2003.

[27] G. Carofiglio, C.-F. Chiasserini, M. Garetto, and E. Leonardi. Route stability in MANETs under the random direction mobility model. *IEEE Transactions on Mobile Computing*, 8(9):1167–1179, 2009.

[28] F. Bai, N. Sadagopan, and A. Helmy. The IMPORTANT framework for analyzing the Impact of Mobility on Performance Of RouTing protocols for Adhoc NeTworks. *Ad Hoc Networks*, 1(4):383–403, 2003.

[29] T. Camp, J. Boleng, and V. Davies. A survey of mobility models for ad hoc network research. *Wireless Communications and Mobile Computing*, 2(5):483–502, 2002.

[30] V. Tolety. *Load reduction in ad hoc networks using mobile servers*, 1999, pp. 1–9.

[31] Y. Wan, K. Namuduri, Y. Zhou, and S. Fu. A smooth-turn mobility model for airborne networks. *IEEE Transactions on Vehicular Technology*, 62(7):3359–3370, 2013.

[32] E. Kuiper and S. Nadjm-Tehrani. Mobility models for UAV group reconnaissance applications. In *2006 International Conference on Wireless and Mobile Communications (ICWMC'06)*, Bucharest, Romania, p. 33. IEEE, 2006.

[33] W. Wang, X. Guan, B. Wang, and Y. Wang. A novel mobility model based on semi-random circular movement in mobile ad hoc networks. *Information Sciences*, 180(3):399–413, 2010.

[34] O. Bouachir, A. Abrassart, F. Garcia, and N. Larrieu. A mobility model for UAV ad hoc network. In *2014 International Conference on Unmanned Aircraft Systems (ICUAS)*, Orlando, FL, pp. 383–388. IEEE, 2014.

[35] D. Broyles, A. Jabbar, and J. P. G. Sterbenz. Design and analysis of a 3-D Gauss-Markov mobility model for highly-dynamic airborne networks. In *Proceedings of the International Telemetering Conference (ITC)*, San Diego, CA, 2010, pp. 25–28.

[36] M. Sánchez and P. Manzoni. ANEJOS: a Java-based simulator for ad hoc networks. *Future Generation Computer Systems*, 17(5):573–583, 2001.

[37] X. Li, T. Zhang, and J. Li. A particle swarm mobility model for flying ad hoc networks. In *GLOBECOM 2017-2017 IEEE Global Communications Conference*, Singapore, pp. 1–6. IEEE, 2017.

[38] J. Sanchez-Garcia, J. M. Garcia-Campos, S. L. Toral, D. G. Reina, and F. Barrero. A self-organizing aerial ad hoc network mobility model for disaster scenarios. In *2015 International Conference on Developments of E-Systems Engineering (DeSE)*, Dubai, pp. 35–40. IEEE, 2015.

[39] N. Shi and X. Luo. A novel cluster-based location-aided routing protocol for UAV fleet networks. *International Journal of Digital Content Technology and Its Applications*, 6(18):376, 2012.

[40] D. Johnson, Y. C. Hu, and D. Maltz. *The Dynamic Source Routing Protocol (DSR) for Mobile Ad Hoc Networks for IPv4*, Vol. 260. RFC Editor, Marina del Rey, CA, 2007.

[41] F. Aadil, A. Raza, M. Khan, M. Maqsood, I. Mehmood, and S. Rho. Energy-aware cluster-based routing in flying ad-hoc networks. *Sensors*, 18(5):1413, 2018.

[42] K. Li, W. Ni, X. Wang, R. P. Liu, S. S. Kanhere, and S. Jha. Energy-efficient cooperative relaying for unmanned aerial vehicles. *IEEE Transactions on Mobile Computing*, 15(6):1377–1386, 2015.

[43] F. Khelifi, A. Bradai, K. Singh, and M. Atri. Localization and energy-efficient data routing for unmanned aerial vehicles: fuzzy-logic-based approach. *IEEE Communications Magazine*, 56(4):129–133, 2018.

[44] J.-A. Maxa, M. S. Ben Mahmoud, and N. Larrieu. Secure routing protocol design for UAV ad hoc networks. In *2015 IEEE/AIAA 34th Digital Avionics Systems Conference (DASC)*, Prague, Czech Republic, pp. 4A5-1–4A5-15. IEEE, 2015.

[45] M. Sbeiti, N. Goddemeier, D. Behnke, and C. Wietfeld. PASER: secure and efficient routing approach for airborne mesh networks. *IEEE Transactions on Wireless Communications*, 15(3):1950–1964, 2015.

[46] J.-A. Maxa, M. S. Ben Mahmoud, and N. Larrieu. Joint model-driven design and real experiment-based validation for a secure UAV ad hoc network routing protocol. In *2016 Integrated Communications Navigation and Surveillance (ICNS)*, Herndon, VA, pp. 1E2-1–1E2-16. IEEE, 2016.

[47] J.-A. Maxa. Model Driven Approach to design a Secure Routing Protocol for UAV Ad Hoc Networks. *EDSYS 2015, 15ème Congrès des doctorants*, EDSYS, May 2015, Toulouse, France. ⟨hal-01166854⟩ V1

[48] A. Aggarwal, S. Gandhi, N. Chaubey, P. Shah, and M. Sadhwani. AODVSEC: a novel approach to secure Ad Hoc on-Demand Distance Vector (AODV) routing protocol from insider attacks in MANETs. *International Journal of Computer Networks & Communications*, 4(4):191–210, 2012.

CHAPTER 10

A Transition from Cloud to Fog Computing: Identifying Features, Challenges and the Future

Namrata Dhanda and Bramah Hazela

CONTENTS

10.1 INTRODUCTION

With data nowadays being generated at an ever-increasing rate every day, efficient computation methods are needed to tackle such large amounts of data. Cloud computing has evolved as a computational utility for handling this data at a nominal cost without the need for huge investment in infrastructure. Developers are able to see their innovations deployed thanks to cloud computing, which has become the preferred choice nowadays due to reduced infrastructure cost, and increased speed, efficiency, productivity, performance and security.

Cloud computing provides scalability, with companies running batch jobs obtaining quicker results running 1000 cloud servers for one hour than running a single server for more than 1000 hours. This resource flexibility provides more efficient and fast computational results at no extra cost.

Cloud computing provides different types of services and resources over the internet, including storage repositories, servers, CPUs, database, application software, and so on. Instead of storing data on a proprietary hard drive or a local storage device, cloud computing enables data storage at a remote location in the cloud. The remote data can be accessed by any electronic device that has web connectivity.

- **Reduced Cost**
  Cloud computing reduces expenditure on hardware and software and on-site data centres – which would include racks of servers, electricity for power and cooling, experts for the management of infrastructure, and so on.

- **Increased Speed**
  Most of the services provided by cloud computing are either self-service or on-demand services. Hence many computing resources can be accessed with just a click of the mouse, providing a lot of flexibility for businesses.

- **Global scaling**
  Cloud computing has the ability to scale elastically: it is capable of delivering the right amount of information or resources like bandwidth, storage and CPU when required.

- **Productivity**
  Rather than investing a lot of time in activities like racking and stacking, hardware setting, software patching, and so on, IT teams can concentrate on more important tasks to achieve their business goals.

- **Performance**
  All cloud computing services run on a worldwide network of secure data centres, which reduces network latency for cloud-based applications compared with a single data centre. These secured networked data centres are updated at regular intervals to the latest versions of fast and efficient computing hardware.

- **Reliability**
  As data can be replicated at a number of sites on a cloud network, cloud computing makes it easier to handle disaster recovery with data backup.

- **Security**
  Cloud computing offers data security by protecting applications from potential threats. A sound and complete set of security policies ensure that data security is not violated.

The various types of services provided by cloud computing can be categorized as follows:

- **IaaS (Infrastructure as a Service)** – This kind of service facilitates a remote data centre and provides resources such as data storage capacity, processing power and networking.

- **PaaS (Platform as a Service)** – Cloud computing provides a platform with different tools and components for the creation, testing and launch of applications.

- **SaaS (Software as a Service)** – Cloud computing also facilitates ready-to-use software to cater to a variety of business requirements.

10.2 FROM CLOUD COMPUTING TO FOG COMPUTING

It has been estimated that by the year 2021 there will be about 25 billion Internet of Things (IoT) devices in the world and this figure is expected to rise to 75 billion by the year 2025 as a large number of devices will be connected through the internet. This device connectivity will generate huge amounts of data that will need to be processed fast and in a sustainable way. In order to meet the increasing demand for IoT solutions, fog computing has evolved, which has certain advantages over cloud computing.

Integrating cloud computing with the IoT is a cost-effective way of increasing business. Off-site services provide the necessary scalability and flexibility to manage and analyze data gathered by connected devices, while specialized platforms (e.g. Azure IoT Suite, IBM Watson, AWS, Google Cloud IoT) give developers the power to create IoT apps without major investment in hardware and software. However, despite the many benefits of cloud computing, there are some areas where it is lacking. The following are some of the drawbacks of cloud computing for IoT:

- **High latency:** A number of IoT applications require very low latency but the cloud environment does not guarantee that because of the distance between the client devices and the data processing centres.

- **Downtime:** Technical issues and interruptions that occur in any internet-based system cause outages for customers. Many companies use multiple connection channels with automated failover to avoid such problems.

- **Security and privacy:** A huge amount of data is transferred through globally connected channels along with thousands of gigabytes of other users' data. Hence, data is vulnerable to cyber-attack or loss. This problem can be resolved with the help of private or hybrid clouds.

10.3 FOG COMPUTING

The term fog computing or "fogging" was coined by Cisco in 2014. Fog and cloud computing are interconnected. Just as fog is closer to the earth than the clouds, in technological terms fog computing is closer to end users than cloud computing. Fog is an extension of cloud computing that consists of a number of edge nodes directly connected to physical devices. These nodes are much closer to the devices than are the data centres [1], enabling them to provide instant connections. The high computational power of the edge nodes allows for instant computation of large amounts of data without the need to send it to distant servers. Fog also includes cloudlets, small-scale powerful data centres located at the edge of the network. The purpose of these cloudlets is to support IoT applications with low latency that require a lot of resources. Fog computing facilitates additional computing, storage and network communication operations between end-user devices and data centres [2–5].

The main purpose of this new technology was to support latency-critical applications such as augmented reality [6] and IoT, which produce massive volumes of data that it is impractical to send to faraway cloud data centres for analysis [7]. There have been various surveys on fog computing and its associated technologies [1, 8, 9]. The main difference between fog computing and cloud computing is that cloud is a centralized system, whereas fog is a distributed decentralized infrastructure. Fog computing acts as an intermediary between the hardware and the remote servers. It decides which information can be locally processed and which information should be sent to the server, thus acting as an intelligent gateway that distributes the cloud load, enabling more efficient data storage, processing and analysis.

10.4 DIFFERENCES BETWEEN CLOUD COMPUTING AND FOG COMPUTING

The following are the differences between cloud and fog computing:

1. Cloud architecture is centralized and consists of large data centres that can be located around the globe, a thousand miles away from client devices. Fog architecture is distributed and consists of millions of small nodes located as close to client devices as possible.

2. Fog acts as a mediator between data centres and hardware, and hence it is closer to end users. If there is no fog layer, the cloud communicates with devices directly, which is time-consuming.

3. In cloud computing, data processing takes place in remote data centres. Fog processing and storage are done on the edge of the network close to the source of information, which is crucial for real-time control.

4. Cloud is more powerful than fog as regards computing capabilities and storage capacity.

5. The cloud consists of a few large server nodes. Fog includes millions of small nodes.

6. Fog performs short-term edge analysis due to its instant responsiveness, while the cloud aims for long-term deep analysis due to its slower responsiveness.

7. Fog provides low latency, whereas cloud provides high latency.

8. A cloud system collapses without an internet connection. Fog computing uses various protocols and standards, so the risk of failure is much lower.

9. Fog is a more secure system than the cloud due to its distributed architecture.

10.5 FOG COMPUTING MODEL

The basic fog model is shown in Figure 10.1. The basic computing components of the fog environment are fog devices, fog servers and gateways. Any computer which has the ability to compute, network and store can act as a fog system. Configuration boxes, switches, routers, base stations, proxy servers or any other network device are part of these systems. Translation services between heterogeneous devices in the fog computer environment are carried out via a fog server which manages multiple fog devices and fog gateways.

In addition, fog gateways offer translation services between IoT, fog and cloud layers. In the past few years, new challenges have arisen in this evolving computing model.

10.5.1 Layered Architecture of Fog Computing

Multi-layered fog machine architecture is shown in Figure 10.2. The components are divided into several classes, which are described as the layer based on their functionality. Such features allow IoT devices to connect with various fog apps, servers, portals and the cloud. This section explains each layer, with the definition providing an example of intelligent transport use.

1. *Physical Layer*: Consists of physical and virtual sensors. Physical sensors include temperature and humidity sensors, CCTV devices, GPS sensors, etc. These are used to monitor traffic signals and can predict future traffic. In addition to physical sensors, the functioning of virtual sensors is also important: when there has been a road

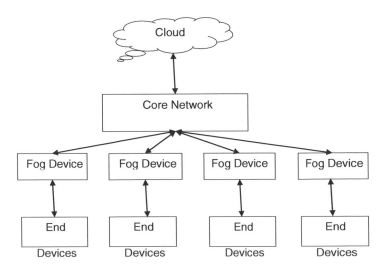

FIGURE 10.1   A fog computing model.

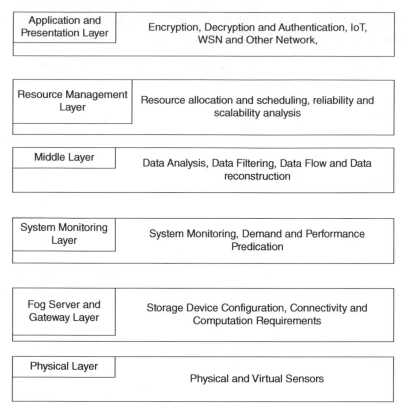

Application and Presentation Layer	Encryption, Decryption and Authentication, IoT, WSN and Other Network,
Resource Management Layer	Resource allocation and scheduling, reliability and scalability analysis
Middle Layer	Data Analysis, Data Filtering, Data Flow and Data reconstruction
System Monitoring Layer	System Monitoring, Demand and Performance Predication
Fog Server and Gateway Layer	Storage Device Configuration, Connectivity and Computation Requirements
Physical Layer	Physical and Virtual Sensors

FIGURE 10.2   Layered fog computing architecture.

accident, a single sensor would not be able to decide whether to block the road or keep traffic moving. The road may have one or more lanes, and another lane will allow the traffic to proceed, but this event reduces traffic handling capacity. The immediate decision on road conditions, traffic multiplexing and traffic reprocessing could be made in this case by a virtual sensor.

2. *Fog Server and Gateway Layer*: This may be a standalone computer or IoT, cloud, fog servers, or fog door. Since the fog server is responsible for several fog devices, it should have greater configured configurations than the fog system and a gateway. There are several factors that make the fog server work. These include its function, its hardware configuration, its networking, its number of devices, etc. Its position determines whether a fog server is separate or part of an IoT system. The fog system will be connected to a community of physical and virtual sensors.

3. *System Monitoring Layer*: Controls system performance and resources. Device management components help select the right tools during service. Similar systems run in smart device environments. It is therefore obvious that when the availability of resources is negative for computation or storage on a fog computer, a problem may occur. The fog server could be a similar case. The fog system and servers need external

help to handle these kinds of situations. The system monitoring component helps to decide such matters efficiently.

4. *Middle Layer*: Includes many components that specifically function on basic and advanced data analysis. Acquired data are processed and extracted at this stage and, if necessary, data trim and reconstruction is also performed. The data flux component will determine whether the data should be stored locally or sent to the cloud for long-lasting storage following processing. In fog computing, the main challenge is storing data at the edge of data and reducing the volume of data to be processed as stream processing. In smart transport systems, for example, data are created from many sensors, and are analyzed and filtered in real time. Any use may be made of all data generated.

5. *Resource Management Layer*: This layer keeps resources allocated and planned, and addresses energy efficiency problems. The reliability of the application schedule and allocation of resources is preserved. Scalability ensures fog services are flexible during peak hours when demand for resources is high. The cloud tackles horizontal scalability and fog calculation strives to give both vertical and horizontal scalability [9]. Network, distribution and storage have many distributed system resources. This is a critical question for distributed services that use request processing. This layer also stores and manages data using virtual storage modules. The data backup component ensures data access and reduces data loss. A network-based pool of storage devices functions as a single, easier-to-maintain storage device within the data virtualization principle.

6. *Application and Presentation Layer*: The components in this layer, which also protect fog users' privacy, will manage any security-related problems, such as communication encryption and protected data storage. Although fog has been developed for IoT applications, fog computing is supported by several other applications based on wireless sensor networks (WSNs) and content delivery networks (CDNs). Any program with latency-conscious characteristics can use fog computing. This includes all kinds of utilities that can fit into fog computing through improved quality and cost efficiency.

For instance, fog computing is by nature an enhanced reality-based application, and so will address the needs of real-time processing for rising reality applications.

10.6 APPLICATIONS OF FOG COMPUTING

Fog computing has huge benefits in real-time applications. It is broadly used in IoT applications that involve real-time data. It acts as an extended version of cloud computing. It is an intermediary between the cloud and end users (it is closer to end users). It can operate either between machine and machine or between human and machine. Fog computing offers several standout advantages over its predecessor, cloud computing. While it utilizes basic cloud computing technology at its core, it addresses several limitations of cloud computing and helps to boost usability and accessibility in different computing environments. The following are the key advantages offered by fog computing:

- Latency Reduction [10–12]
- Maximizes network bandwidth utilization [13–14]
- Computational load balancing [15]
- Enhanced security [16]
- More energy efficient [17]
- Content delivery network [18]
- Monitoring of edge devices [19]
- Globally distributed network helps keep downtime to a minimum
- Optimal operational expense
- Business agility
- Improved service management

Fog computing applications include the following:

a. **Mobile Big-Data Analytics in IoT.** Data is collected in bulk and cannot be stored in the cloud. In these situations it is beneficial to use fog computing instead of cloud computing as fog nodes are much closer to end systems. It also eliminates other problems such as delay, traffic, processing speed, delivery time, response time, and data storage, transportation and processing. Fog computing could be the future of IoT applications.

b. **Water Pressure Sensors at Dams.** Sensors installed in dams send data to the cloud, where it is analyzed, and officials are alerted in case of any discrepancies. The problem here is that any delay to the information could be dangerous. The solution is fog, which is near the end systems, so can more easily transmit data, analyze it and give instantaneous feedback.

c. **Smart Utility Service.** Here the main aim is to conserve time, money and energy. Data analysis needs to run every minute to remain up to date. This mostly involves the end users, so cloud may not be suitable. Every day these applications provide users with information as to which of their appliances is conserving more energy. IoT also creates a lot of traffic in networks where it is difficult to send other data; therefore fog is a good alternative.

d. **Health Data.** When data needs to be transferred from one hospital to another, high security and data integrity is a must. This can be provided by fog since the data are transferred locally. These fog nodes can be used to update the patient's lab records which can then easily be accessed by local hospitals. Hard copies of patients' medical history need not be kept as these unified records can be accessed by any doctor.

e. **Smart cities.** These are rapidly evolving as a result of technological innovation and IoT proliferation in nearly every aspect of urban life. Ten years ago, the number of sensors embedded throughout a city was small, and now they are almost everywhere – from roads to thermostats and even rubbish bins. It was the ability to connect, communicate and remotely manage different devices that gave rise to the new trend of fog computing.

10.7 ADVANTAGES OF FOG COMPUTING

One of the challenges that smart cities face is the wealth of information that is generated, captured and analyzed every day. As the number of IoT devices is increasing, the amount of collected data becomes overwhelming and requires heavy-duty computing resources to process it. It is also time-consuming and expensive to transfer the large volume of data between the cloud and data sources. Fog computing makes it possible to reduce the amount of data that needs to be sent to the cloud for processing, improving efficiency. There are huge benefits of fog computing for smart cities:

- **A minimal amount of data sent to the cloud**

   The key goal of fog computing is to make big data smaller. It is estimated that the volume of data captured by connected devices will exceed 79.4 zettabytes by 2025. Fog computing is capable of reducing the amount of data by applying intelligent sensing and filtering, which allows only useful information to be transmitted, based on knowledge available locally at a given fog device. Data analytics further minimizes data volumes by throwing away raw data.

- **Low data latency**

   Fog nodes can process data on board without sending it to remote cloud servers and delivering the results back. This reduces data travel time considerably and enables responses to be received in real time. Immediate data processing might be critical for smart-city systems, especially when actions need to be taken quickly: for example, turning traffic lights to green when emergency vehicles with lights and sirens are approaching might save lives.

- **Reduced bandwidth**

   Transmitting and processing data requires massive bandwidth, which in cloud computing is limited. However, this is not an issue with fog computing, since all data is distributed between local devices and not sent wirelessly, allowing for a significant decrease in network bandwidth consumption.

- **Enhanced data security**

   Data security is another major reason smart cities are turning to fog computing. It keeps sensitive and confidential data away from vulnerable public networks, preventing cybercriminals from gaining access to it. With fog computing, malware and infected files can be found at an early stage at device level before they infect the whole network.

10.8 APPLICATION OF FOG COMPUTING IN SMART CITIES

Fog computing has great potential to become the next big trend in smart cities due to its ability to swiftly and securely process small volumes of data. It helps collect data on city activities from road traffic and public safety to waste management and air quality, ensuring everything is running efficiently and bringing sustainability to urban life. With the aid of fog computing, the data can be quickly processed and analyzed in order to run systems more effectively. However, the fog-only approach is not able to deal with huge amounts of data, so the cloud will continue to have a role in the IoT ecosystem.

10.8.1 Traffic Regulation

Smart cities use different types of sensors to monitor and regulate road traffic. Sensors embedded in smart traffic lights detect passing pedestrians, cyclists and drivers; measure their speed and the distance between them; analyze traffic data as they collect it; and take real-time decisions to alter the lights or reroute part of the traffic if necessary. This contributes to improved traffic flow, and fewer road accidents and casualties. It also makes it possible for drivers to rely on autonomous vehicles in an emergency. The collected data might be later sent to the cloud for long-term analysis.

Fog computing combined with smart traffic lights has already proved very effective in dealing with traffic congestion. As an example, a community in Bellevue, Washington installed intelligent traffic lights that respond to traffic conditions in real time: the green lights stay on longer during peak traffic periods. City officials estimated that this led to a 36 per cent reduction in travel time on the city's main road, saving drivers US$9 to12 million annually.

Palo Alto, California is another city that launched a smart traffic signal project enabling the integration of traffic lights with connected vehicles. The project started in 2016 and is expected to improve traffic signal timing considerably.

10.8.2 Waste Management

Waste management often represents a challenge to smart cities, since it is a process that requires a lot of time, money and resources. Refuse is collected from containers on particular days according to a schedule, without taking into account how much waste is in them. Collecting waste from a nearly empty container is not efficient as it leads to unnecessary fuel consumption, whereas overfilled bins make the streets look dirty.

The application of smart sensors and fog computing will allow real-time monitoring of refuse levels and provide a way for more efficient waste management. Sensors installed on rubbish bins could identify that they are almost full and alert the refuse collection department. Fill-level data could be sent to the cloud for long-range analysis in order to optimize routes and schedules for the refuse collection trucks.

10.8.3 Environmental Control

Environmental parameters concerning the city's natural resources can be monitored and analyzed with fog computing. For example, a smart water system can be expected to analyze water quality and detect any deviation from the norm, such as high nitrate or iron levels. Furthermore, it can detect leaks and immediately inform maintenance teams.

Greenhouse gas control is another area to which connected technology and fog computing might be applied to improve environmental sustainability. The analysis of actionable data enables a city government to see the overall picture of greenhouse gas emissions and take remedial action in time. Based on the results, they might send citizens reminders to use less heating or hot water in order to help reduce greenhouse gas emissions.

10.8.4 Surveillance Systems

Video surveillance systems equipped with smart sensors make a major contribution to security and safety on city streets. The large amount of information generated needs be analyzed in real time in order to ensure effective monitoring of public spaces. Traditional cloud-based models are not suitable for these purposes due to the massive amount of data, associated latency challenges, network availability and the enormous expense required to continually stream the data to the cloud and back. This is where fog computing comes in.

With fog computing, data collected from video surveillance cameras is stored and processed on fog nodes close to the edge. Low latency allows for effective surveillance and detection of anti-social behaviour in public spaces, such as a busy airport or a shopping centre. Once an incident happens, law-enforcement authorities or security services will receive an alert allowing them to act quickly.

10.9 LIMITATIONS OF ADOPTING FOG COMPUTING

Some hurdles are encountered in the move from cloud to fog computing, and these are listed below:

1. No Edge Services:

   With its services-based consumption model, the cloud is easily available, yet there are currently no specialized network facilities where on-demand computing or storage space can be offered. We expect that sooner or later, even with fog services in partnership with cloud providers and network operators, management of leading networks will be on the market. Amazon's Greengrass service, which provides edge-device technology, is taking the first steps in this direction. Nonetheless, since there is currently no way to really 'share' the edge capabilities of the fog application provider, its own physical edge boxes, including cloud integration, must be installed, managed and run now.

2. Standardized equipment shortage:

   In terms of computing power, Raspberry Pi, Beagle Boards or customized solutions are also available. It leads to a large heterogeneous hardware edge node that has two effects: First, software stacks must be modified or not run anywhere. Systems which are actually capable of running everywhere will not take full advantage of their ability. Second, the architectures of applications need to be able to handle these diverse resources. It includes highly dynamic systems that can provide some or all of the service functions based on the nodes in which they operate.

3. Effort Management:

As it is primarily about data and computation getting the end users closer to fog computing, the dense deployment of fog can lead to very large numbers of fog nodes. All of these must be handled, for example, when scales are made, upgrades installed or settings updated. As the fog infrastructure service is not currently managed, this management effort is left to those who care to do it, a much greater effort than for tiny on-site data centres.

4. Network Transparency:

Distributed systems work also attempted to hide deployment, with programs operating as though they were running on a single machine. More and more transparencies in favour of other QoS targets have had to be discarded over the years. In fog computing, however, the interconnection between nodes at a logical level can no longer be arbitrary, because, in reality, the underlying networks are more or less hierarchical, i.e., two nodes at the geographic edge can be directly connected on the next highest hierarchical level, at or between the internet backbones.

5. Physical Security:

Traditional data centres, including but not limited to the in-situ security station, may employ an arbitrary complex physical access control step. This is usually audited and certified to ensure that their measures of physical security are equal. This cannot be achieved in the case of fog nodes that are widely spread over a hostile environment. Physical safety could here mean that the edge box of a streetlight can be mounted on top of the pole (instead of at eye level) [10].

6. Regulatory Requirements:

There are legal and regulatory provisions for the preservation of data in certain physical locations in many countries, for example in the field of safety. In addition, storage and treatment plants must meet those criteria demonstrably. Cloud data centres that operate in each country can be accredited, but for edge nodes it tends to be impractical. In addition, privacy law, including the EU's GDPR [9], also covers elements such as transparency and the right of people to know where sensitive information about them is stored. In liquid fog-based applications, this is challenging to prove as the application may not even know the data location due to virtualization.

REFERENCES

[1] A. Yousefpour, C. Fung, T. Nguyen, K. Kadiyala, F. Jalali, A. Niakanlahiji, J. Kong, and J. P. Jue. All one needs to know about fog computing and related edge computing paradigms: a complete survey. *Journal of Systems Architecture*, 98:289–330, 2019.

[2] F. Bonomi, R. Milito, J. Zhu, and S. Addepalli. Fog computing and its role in the Internet of Things. In *Proceedings of the Workshop on Mobile Computing*, Helsinki, Finland, 2012.

[3] L. Vaquero. Finding your way in the fog: towards a comprehensive definition of fog computing. *ACM SIGCOMM Computer Communication Review*, 44(5):27–32, 2014.

[4] OpenFog Consortium. OpenFog reference architecture for fog computing. *Architecture Working Group*. 2017. Feb:1–62. Available at: https://www.openfogconsortium.org/ra/.

[5] IEEE Standards Association. IEEE 1934-2018 – IEEE standard for adoption of OpenFog reference architecture for fog computing. 2018. Availablet at: https://standards.ieee.org/standard/1934-2018.html.

[6] X. Chen, Decentralized computation offloading game for mobile cloud computing. *IEEE Transactions on Parallel and Distributed Systems*, 26(4):974–983, 2014.

[7] H. Atlam, R. Walters, and G. Wills. Fog computing and the Internet of Things: a review. *Big Data and Cognitive Computing*, 2(2): 10–28, 2018.

[8] R. Mahmud, R. Kotagiri, and R. Buyya. *Fog Computing: A Taxonomy, Survey, and Future Directions*. Springer, Singapore, 2018.

[9] R. K. Naha, S. K. Garg, D. Georgekopolous, P. P. Jayaraman, L. Gao, Y. Xiang, and R. Ranjan. Fog computing: survey of trends, architectures, requirements, and research directions. *IEEE Access*, 6, 47980–48009, 2018. http://arxiv.org/abs/1807.00976.

[10] M. Abrash. Latency – the sine quanon of AR and VR. *Blog post*, Dec 2012. Available at: http://blogs.valvesoftware.com/abrash/latency-the-sine-qua-non-of-ar-and-vr/.

[11] OculusRift. Delivers some home truths on latency. Oculus VR, Menlo Park, CA, Available at: https://oculusrift-blog.com/john-carmacks-message-of-latency/682/. Accessed 15 July 2016.

[12] O. Tomanek, and L. Kencl Claudit. Planetary-scale cloud latency auditing platforms. In *IEEE 2nd International Conference on Cloud Networking (CloudNet)*, pp. 138–146. IEEE, 2013. Available at: http://bit.do/bS4js.

[13] T. Stack. Internet of Things (IoT) data continues to explode exponentially. Who is using that data, and how? 2019. Cisco Blogs 5 (2018). Available at: https://blogs.cisco.com/datacenter/internet-of-things-iot-data-continues-to-explode-exponentially-who-is-using-that-data-and-how.

[14] M. R. Shahid, G. Blanc, Z. Zhang, and H. Debar. IoT devices recognition through network traffic analysis. In *Proceedings of the Big Data*, Seattle, WA, December 2018.

[15] J. Oueis, E. C. Strinati, and S. Barbarossa. The fog balancing: Load distribution for small cell cloud computing. In *Proceedings of the VTC*, 2015.

[16] S. J. Stolfo, M. B. Salem, and A. D. Keromytis. Fog computing: mitigating insider data theft attacks in the cloud. In *Proceedings of the IEEE Symposium on Security and Privacy Workshops*, San Francisco, CA, 2012.

[17] F. Jalali, A. Vishwanath, J. De Hoog, and F. Suits. Interconnecting fog computing and microgrids for greening IoT. In *Proceedings of the ISGT-Asia*, Melbourne, Australia, 2016.

[18] X. Wang, S. Leng, and K. Yang. Social-aware edge caching in fog radio access networks. *IEEE Access*, 5:8492–8501, 2017.

[19] OpenFog Consortium. *Patient monitoring*. 2019. https://www.openfogconsortium.org/wp-content/uploads/Patient-Monitoring-Short.pdf.

Index

Page numbers in *italic* indicate a figure and page numbers in **bold** indicate a table on the corresponding page.